Energy and the Environment

Liberty and the Environment

Special Publication No.81

Energy and the Environment

The Proceedings of a Symposium organized jointly
by the Inorganic Chemicals Group and the
Environment Group of the Industrial Division of
the Royal Society of Chemistry

University of Leeds, 3rd–5th April 1990

Edited by
J Dunderdale

ROYAL
SOCIETY OF
CHEMISTRY

British Library Cataloguing in Publication Data

Energy and the environment: the proceedings of a symposium
  organized jointly by the Inorganic Chemicals Group and the
  Environment Group of the Industrial Division of the Royal
  Society of Chemistry.
  1. Environment. Pollution
  I. Dunderdale, J. II. Royal Society of Chemistry III. Series
  363.73

ISBN 0-85186-647-6

Published by the Royal Society of Chemistry, Thomas Graham House, Science Park,
Cambridge CB4 4WF

Printed in Great Britain by Whitstable Litho Ltd, Whitstable, Kent

Preface

It has been said that Global Warming is the second greatest problem facing mankind today. It may not be long before it assumes the major role. In recent years there have been a number of unusual, violent, localised weather disturbances which many attribute to wider than normal excursions in the average weather pattern; there have also been changes in regional weather patterns on a global basis. Are these variations just statistical fluctuations or are they indeed early symptoms of the 'Greenhouse Effect'? Unless it can be positively stated that the observed phenomena are due to statistical fluctuations only, then, as Professor Fells points out in the opening paper, ".... an immediate coordinated attack is required". Improved efficiency of production and use may reduce the rate of onset but in itself is insufficient, other alternatives for reducing carbon dioxide emissions are not being adopted because of a widespread lack of knowledge of the relative risks to the biosphere from the options available.

The conference was arranged with the above thoughts in mind; the purpose being to consider the relative environmental impact of each step in each of the major energy producing processes. The opening paper relates energy consumption and consequent global warming to population growth and improved lifestyle; the need for education and immediate action is stressed. The remaining papers fall naturally into groups; the first group (papers 2 - 5) deals with the chemical interactions of gaseous and vaporous emissions, from all energy producing and using processes, in and with the atmosphere and biosphere; the second group (papers 6 to 10) looks in detail at the pollutants arising and the control measures adopted at each stage in the production of nuclear power from mining to generation - including waste disposal, fuel reprocessing, accidental releases and decommissioning - while the final paper on nuclear topics provides a survey of the fusion process; papers 11 - 13 cover fossil fuel processes; the next group (papers 14 - 16) discusses the environmental legislation (UK & EEC) which applies to energy producers and users and the means taken to meet these requirements; finally there are two papers which consider conservation and alternative energy sources.

The conference was also arranged with a view to publishing as there appeared to be a need for a collected data source which would provide the reader with the opportunity to compare the environmental impact of the various energy producing and using processes. The quoted legislation relating to emission limits, against which performance can be assessed, is essentially European but is comparable with other existing requirements. Apart from some brief, technical, passages, the book should be intelligible to a wide audience.

The RSC would like to express its appreciation to the various organisations and departments for their permission to publish the papers herein, and on behalf of the other members of the organising committee and myself I would like to thank the authors for their contributions and their cooperation in producing their papers in a form suitable for publication.

John Dunderdale

Inorganic Chemicals Group

Contents

The Effects of Pollutants on Materials
R. N. Butlin and T. J. S. Yates

The Environmental Impact of the Nuclear Fuel Cycle
L. E. J. Roberts

**The Generation, Storage, Treatment and Disposal of
Nuclear Waste at Sellafield**
L. F. Johnson and W. F. Larkins

Contents

Severe Accidents
P. N. Clough

**The Potential Radiological Consequences of Deferring
the Final Dismantling of a Magnox Nuclear Power Station**
P. Woollam

Fusion Reactors and the Environment
R. Hancox

**The Disposal of Solid and Liquid Wastes from Coal
Washing Operations**
D. W. Brown

**The Disposal of Solid Combustion Products from Power
Stations**
*B. H. M. Billinge, A. F. Dillon, F. Harrison and
D. Tidy*

Contents

The Environmental Effects of Oil and Gas Production
D. E. Martin

Vehicle Emission Control Technology
J. M. Dunne

Control of Emissions from Stationary Sources of Fossil Fuel Combustion
M. J. Cooke and R. J. Pragnell

Emission Control - Statutory Requirements
K. Speakman

The Problem

I. Fells
DEPARTMENT OF CHEMICAL AND PROCESS ENGINEERING, THE
UNIVERSITY OF NEWCASTLE-UPON-TYNE, MERZ COURT, CLAREMONT ROAD,
NEWCASTLE-UPON-TYNE NE1 7RU, UK

THE PROBLEM

Energy is the convertible currency of technology. Without energy the whole fabric of society as we know it would crumble; the effect of a 24 hour cut in electricity supplies to a city shows how totally dependent we are on that particularly useful form of energy. Computers and lifts cease to function, hospitals sink to a care and maintenance level and the lights go out. As populations grow, many faster than the average 2 per cent, need for more and more energy is exacerbated. Enhanced lifestyle and energy demand rise together and the wealthy industrialised economies which contain 25 per cent of the world's population consume 75 per cent of the world's energy supply.

Unfortunately this insatiable desire for more and more energy can only be met by burning the world's reserves of fossil fuels or wood, that invaluable but threatened renewable resource. Ninety per cent of world energy is provided by burning one hydrocarbon fuel or another and the inescapable result is to add more carbon dioxide to the already burdened atmosphere and so increase global warming with its attendant destabilising effect on the weather machine. Worldwide we burn 5.5 billion tonnes of carbon each year which produces more than 20 billion tonnes of carbon dioxide. Half of this carbon dioxide is absorbed into sinks of various kinds; the sea, vegetation and so on but half remains to add to carbon dioxide in the atmosphere which has been steadily accumulating since the industrial revolution started. The insulating effect of this carbon dioxide blanket is causing the world to warm up!

Acid rain, largely arising from the combustion of fossil fuel for electricity generation is also contributing to the degradation of our environment. Dying forests, sterile lakes and corroded buildings are all now bearing witness to the folly of unrestrained combustion of our fuel reserves.

EXPECTATIONS

Demand for fuel has risen steadily since the mid-19th century at a rate rising from 2 per cent to over 4 per cent in the 1960s. It flattened out through the 1970s in the developed industrial world, where a successful move away from oil to coal and nuclear power took place, but overall it rose by 3.7 per cent in '84 as the world came out of recession and is now rising at 2.8 per cent per annum. However, the fast growing economies of the developing world, some rising at 8 per cent per annum and more, and with energy coefficients of 1.5 (that is a 1.5 per cent increase in energy demand for every 1 per cent increase in GDP), had to buy oil to sustain their industrial growth. The second OPEC-engendered enormous oil price increase of 1979, which drove the price to $40 per barrel, led to a Third World debt which now approaches $1000bn. There seems little chance of this ever being repaid and the failure of the world's banking system to cope with petrodollars is regrettable. The high and volatile cost of oil led Mrs Gandhi to announce in 1983, at the World Energy Conference in India, a nuclear programme of 10,000 MW based on home designed and built nuclear power stations to be completed by the year 2000. To those who complained of this proliferation, wishing to confine nuclear power to the industrialised nations, she pointed to the modest expectation of economic growth and lifestyles of their 700,000,000 people (1000,000,000 before the end of the century) which would be unrealisable without home-produced affordable energy. Any risk from possible nuclear accidents paled into insignificance beside the risk of political instability which would arise if she were able to give no assurance of an improved future life style.

The demand for energy by the developing world will place the biggest strain on world energy resources. Lacking the technology to generate and distribute electricity extensively, many developing countries rely on oil, although this is changing as coal and nuclear power expand. Sadly, the one important renewable resource, fuel wood, which supplies 20 or even 30 per cent of energy in some rural economies, is running out and there is a world shortage. It is ironic that the first fuel resource to run into really short supply will be a 'renewable' resource. The effect of this shortage on hundreds of millions of inhabitants of sub-Saharan Africa is to reduce even further their subsistence level lifestyle. They are the real 'have nots'.

If the whole world could be brought up to Western Europe's living standard, which requires for its maintenance about 6 tonnes of coal equivalent per person per year (that is 6 1kW electric fires burning day and night for each of us) then the world's proved fossil fuel resources, coal, oil and gas, would last for just 17 short years. As it is, there are 31.4×10^{21} Joules of proved renewable fossil fuels reserves (coal 20.3, oil and gas 11.1); at current rates of demand of 400×10^{18} Joules per annum the resources will last for 80 years. Nuclear energy could transform the situation if we choose to use it. As it is, every year we burn fossil fuel that took a million years to form so that in some 200 years, starting in the middle of the last century, we will have burnt all the fossil fuel formed during the history of the world. There is currently a glut of fossil fuels but as the developing world industrialises and its economies expand, that glut could quickly become a famine in the second decade of the next century (25 years away).

One solution to the impending energy shortage is to make much more use of renewable energy sources. This cause is sometimes espoused with a fervour which leads to extravagant and impossible claims being made. Engineering practicality, costs, scarcity of supply and public acceptability must all be considered.

Ultimately, of course, all energy supplies on earth derive from the sun and solar radiant energy provides a continuous stream of energy which warms us, causes crops to grow via photosynthesis, heats the land and sea differentially and so causes winds and consequently waves and, of course, rain leading to hydro power. Tidal rise and fall is the result of gravitational pull of moon and sun and geothermal heat the result of radioactive decay deep in the earth. All are possible sources of energy but though the science is understood, it does not follow that provided enough research money is poured into the project

an engineering solution will be found.   The scientific
understanding of a process is the easy part;. it is the
engineering that is difficult.

     In the case of wave power, for example, although the
average power in a metre length of North Sea wave is 50 kW
it turns out to be so difficult to build a device that
will both tap this energy source and stand up to the
rigours of the North Sea environment that it can only be
made so robust that the electricity generated costs three
times what it does when generated from coal or oil.   In
Israel, where enormous effort has been put into trapping
and using solar energy in what is an extremely favourable
solar climate, some 20 years into the programme only 3 per
cent of Israel's energy comes from solar power.   Wind
power is seen as a better prospect than solar power in the
United Kingdom but the engineering is difficult because
the power varies as the cube of the velocity so that at
low wind speeds electricity generation is very inefficient
and at high speeds the wind generator must close down to
avoid damage.   Nevertheless it is claimed that electricity
can be generated for the same price as from coal, and wind
generators are being built to feed into the grid.   Some
indication of their potential can be gained from Danish
experience where the windy landscape is dotted with wind
generators of around 200 kW output built with a generous
subsidy.   They provide only 0.3 per cent of Danish
electricity and it is thought the maximum they could
provide will be 2 to 3 per cent of demand. World wide,
hydropower provides about 7 per cent of total energy
demand, but in the UK it provides only 1 per cent of
electricity because of the paucity of the right kind of
geographical location.

     Perhaps the best renewable prospect in the UK is the
Severn Barrage which would harness the high (10m) rise and
fall of the tide in the Severn Estuary to provide 7.2 GW
of peak power and steady power of around 2 GW, which is
equivalent to 2 nuclear power stations, at around the same
cost of generation as nuclear or coal-fired power.   The
capital cost of the venture would be around £8,000 million
and there are some worries that it would intrude on the
environment, upsetting the habitat of wading birds and the
like.   Indeed, the environmental impact of renewable
energy sources, whether it be crops for fuel, wave power
changing the shoreline, wind power, hydropower, is just as
profound as for traditional energy sources.   Because of
the diffuse nature of many renewable resources their
exploitation would neutralise far greater ground area than

traditional systems. Drax coal-fired power station, for example, generates electricity at a rate of 4000 MW using 425 acres of land. The same area covered with large 3 MW wind generators would supply 100 MW, just one fortieth of the coal station output.

A possible source of energy, which is often spoken of in the same breath as renewables, is fusion. The process, whereby 2 isotopic atoms of hydrogen are fused at 100 million degrees K to give a helium atom plus a massive burp of energy, has been demonstrated in the hydrogen bomb, but to control the fusion process on a large enough scale to construct a power station around it has not yet been achieved, and the engineering of such a station might prove just too difficult. In any event such a development, if successful, could not generate more than 2 or 3 per cent of world electricity in the next 50 years. Such a scheme would carry a radioactive waste penalty, though not as extensive as from nuclear fission stations.

Of course, coal is one major contender as a power supply; there is some 20 times more recoverable coal than oil and gas in the world. It comes from potentially more stable areas than the Middle East and the technology is well understood and mature. Coal consumption is growing by 3 per cent per annum and world trade in coal is expanding at 6 per cent per annum. China, for example, adds 40m tonnes of output each year. Coal provided some 80 per cent of UK electricity and world wide, taken with nuclear power, exerts a vital stabilising influence on world energy supply in contrast to the highly economically and politically volatile oil. (Gas tends to be linked to oil in terms of price and often volatility.)

Nuclear fission presents paradoxically the biggest potential and the biggest problem. The second generation proven fast breeder reactor system, which uses uranium fuel 60 times more efficiently than today's ubiquitous thermal nuclear reactors, could supply electricity for the world's needs for the next 500 years and more, if we are prepared to use it. Today's thermal reactors provide 20 per cent of world electricity; there are 433 reactors in 26 countries generating more than 300,000 MW of power and a further 98 reactors are under construction.

Tragically, on April 26th 1986 the fourth unit in the RBMK nuclear power station at Chernobyl in the Ukraine suffered a catastrophic accident which resulted in the destruction of the reactor core and part of the building.

Radioactive debris was flung 2 km into the atmosphere and drifted across Scandinavian and Western Europe leading to contamination of the food chain by radioactive iodine and, more importantly, the relatively stable caesium 137 with a half life of 30 years. Even now some lamb and reindeer meat is considered unsafe to eat. In the UK the average dose of radiation received by the population of the UK as a result of the release is 0.1 mSv or about 5 per cent of our background dose for one year. The overall effect of Chernobyl will be to increase the cancer mortality in the northern hemisphere by 0.0025 per cent. One of the many lessons learnt from this tragic accident, where once again the fallibility of the human operators played an important role in causing the accident, is the universality of the results of the accident. Radioactive debris is no respecter of national boundaries and one country's accident rapidly becomes the world's accident. The effect of Chernobyl has been to cause shock and dismay in many countries, and a reaffirmation in countries such as Sweden to phase out nuclear power and in Denmark not to install it. New nuclear power station ordering in the USA had already ceased following the accident at Three Mile Island where, mercifully, no one died. But in other parts of the world Chernobyl has merely been regretted and seen as a series of lessons to be learned. In Japan they will complete two new reactors per year up to 2010 and three per year thereafter. The International Atomic Energy Agency is insisting on a series of further safety improvements in civil nuclear reactors. Curiously and unexpectedly the atmosphere at a meeting of the world nuclear industry in Cannes in October 1986 was one of confidence in that the unthinkable nuclear accident had occurred and the world seemed not to have changed very much. A similar reaction to that of energy economists in the autumn of 1973 after the oil price rise; but Hans Blick of IAEA warned in Cannes that the nuclear future was fragile; one more major nuclear accident anywhere in the world might slow down nuclear growth irrevocably. Caution in the continuing rapid development of nuclear power until new inherently safe reactors have been developed was urged by the Russian delegate Beschinski at the World Energy Conference in 1989.

POLLUTION OF THE ENVIRONMENT

Since the beginning of the industrial revolution the world has indulged in an orgy of combustion, burning fossil fuels, firewood and rain forests indiscriminately. Unfortunately any combustion process involving a

hydrocarbon forms carbon dioxide as a product. Some of
the carbon dioxide is dissolved in the oceans, some is
taken up by vegetation of all kinds including the biomass
of the sea as well as plants and forests, but about half
of the 20 thousand million tonnes of carbon dioxide
produced each year makes its way into the upper
atmosphere. There it traps the infra-red radiation which
would otherwise escape, affecting the process that keeps
the earth in equilibrium with the shorter wave length
incident radiation it receives from the sun and causing a
gradual global warming. Measurement of carbon dioxide
concentrations in bubbles of air trapped in the Antarctic
ice have shown that for 10,000 years it remained constant
at 270 ppm and only started to rise 100 years ago. By
1957 it was 315 ppm and now it is 350 ppm. The average
global warming of $0.5^{\circ}C$ or so masks considerable
fluctuations in different geographical locations but the
trend is seen as unmistakable. Other gases add to the
greenhouse effect of carbon dioxide. Methane arising from
agricultural animals, paddy fields and oil and gas wells
is increasing by 1.2 per cent per year,
chlorofluorocarbons are particularly efficient greenhouse
gases and nitrous oxide and ozone all taken together give
a warming effect as great as carbon dioxide. Computer
models suggest an overall warming of $2^{\circ}C$ will occur by
2030. The effect will be to materially alter climate in
different parts of the world in ways which are as yet
unpredictable. There are already suggestions that the
'weather machine' is becoming increasingly unstable and
that the desertification of Central Africa is already a
result of global warming although the warming effect is
likely to be greatest near the poles. Sea levels will
also rise; they have risen by 150 mm this century.

Another more tractable effect of burning fossil
fuels, but particularly coal and oil, is 'acid rain'.
These fuels particularly contain sulphur and some organic
nitrogen which turns into sulphur dioxide and nitrogen
oxide during combustion and are emitted from power station
chimneys. They undergo a range of chemical and physical
processes depending upon meteorological conditions,
leading to acid deposition at ground level and also the
formation of ozone. Deposition of acid rain over 1000 km
from the source is not unusual and Scandinavian countries
complain that at least half the acid pollution of their
lakes originated in the United Kingdom, where 250kg of
sulphuric acid is put into the atmosphere each year for
every man, woman and child in the population. Acid
emissions can be removed in the case of sulphur oxides by

absorption by calcium oxide or bisulphite solution. This can lead to a serious problem of disposal of the resultant waste, which is in the form of many thousands of tonnes of calcium sulphate or lesser quantities of elemental sulphur. The capital cost is considerable being around £200 x 10^6 for a station of 2,000 MW capacity; this will put 10 per cent on the cost of the electricity produced. The equipment is also more susceptible to 'outage' and the efficiency of conversion of the station is reduced from around 38 per cent from the very best to 36 per cent. Nitrogen oxides are minimised by correct burner design. The process increases the production of CO_2 per unit of electricity generated so worsening the greenhouse effect.

In principle carbon dioxide could be scrubbed out of combustion gases but the cost would be enormous, doubling the cost of the electricity.

Transport, using gasoline or diesel fuel, contributes a quarter of total annual carbon dioxide emissions. Much of this fuel is used very inefficiently in poorly-maintained engines often operating in congested urban conditions. Thermodynamically it is very much better to use diesel rather than gasoline engines. The use of unleaded or 'green' gasoline causes engines to run less efficiently and consequently produce more carbon dioxide than if high octane leaded petrol were used.

If carbon dioxide emission from transport systems is to be reduced a move to large scale, efficient public transport systems, particularly in industrialised countries such as USA, UK and Western Europe, is essential. Electric traction for essential urban transport systems, freight delivery, commuter cars encouraged by legislation could provoke motor manufacturers to take electric traction seriously. And that anachronism, the company car, should be done away with.

EDUCATION

Renewable energy resources all carry environmental penalty of one kind or another. Nuclear power, provided the chemical engineering is of the highest standard, may come to be seen as having the least environmental impact provided public perception of the technology is not clouded by a confused emotional response. This means a better public understanding of science.

The problem for many people, and political decision makers in particular in the UK and elsewhere, is to know what decision to come to about nuclear power. It is easy to say "No!" to it but is that the correct, responsible decision? The nuclear industry, born out of weaponry has always maintained a secretive attitude towards explaining its activities and particularly its mistakes. The effect on the population of this attitude has been to erode confidence in the industry and attract story-seeking investigative journalists who have produced some extremely damaging, anxiety-inducing programmes about nuclear power. Trying to defend itself, the industry has turned to probabilistic risk assessment to try to put it into perspective alongside other risks. Quite a good method is used by Bernard Cohen in his book 'Before it's too late: a scientist's case for nuclear energy'. He lists loss of life expectancy (LLE) in days due to various risks, some avoidable and some not. Here is a selection:

Being male rather than female	2,800
Heart disease	2,100
Being unmarried	2,000
Cigarettes (1 packet per day)	1,600
Cancer	980
Stroke	520
15lbs overweight	450
Motor vehicle accidents	200
Alcohol	130
Small cars v. standard size	50
Firearms	11
All electric power in US nuclear (calculated by the anti-nuclear Society of Concerned Scientists)	1.5
Airline crashes	1
All electric power in US nuclear (calculated by Nuclear Regulatory Commission)	0.03

Looking at the list one might imagine that people had a good deal more to worry about than nuclear power. But this kind of balanced view is not how most people come to decisions. The unlikely chance of a terrifying accident is not seen as remote or acceptable; after all, people do win the football pools despite the long odds. People form their opinions on information gleaned from a wide variety of sources but their problem is a lack of any kind of basic scientific education which can help them reject what is scientific nonsense. This is compounded by the disdain for technology inculcated by our education system in those that become senior civil servants and read Greek, History or English at Oxford and elsewhere. The

French Civil Service is stocked with people educated in
science and engineering at the Ecole Polytechnique and
Ecole Normale Superieure and their enthusiasm and
understanding of science is very different from ours. The
man in the street is particularly badly served by the
education system, where almost no science is taught at the
primary and junior level other than botany and the like,
usually by teachers not trained in science.   Chemistry
and mathematics are particularly badly taught with a few
exceptions. Journalists make the most elementary mistakes
in science, if they try to use it at all, when explaining
technology or medical stories, hence perhaps the general
enthusiasm for holistic or alternative medicine.

        One definition of education is "to enable one to
withstand deleterious propaganda".   It is a deficiency in
this aspect of scientific education which leaves people in
general ill-informed and prejudiced as a result of relying
on a scientifically illiterate media. The faults are not,
however, all on one side.   In its report 'The Public
Understanding of Science', the Royal Society lays a
responsibility on each professional scientist   (and
engineer) to promote public understanding of science: in
the past it has never before been considered a
professional requirement for a scientist to explain in
layman's terms what he is doing, even though his work may
be funded by public money.   It will be difficult to
change, particularly for the nuclear industry, and it
would make the task easier if the public in general were
scientifically literate.   An excellent example of public
ignorance is the emotive subject of radiation. Few people
would be prepared to be in the same room as a new uranium
fuel assembly or know that plutonium, which is certainly
not the most dangerous material known to man, can be
safely, if cautiously, handled encased only in a plastic
sachet.   Information on radiation is available, if
somewhat jargonised; but those vociferous, concerned,
scientifically illiterate, successful, influential trend
setters, who have some sort of affinity with what is
natural and have never bothered to find out the facts,
have an altogether disproportionate influence on MPs and
the like, who often share their characteristics.   It is
little wonder therefore that Nigel Lawson, when he was
Secretary of State for Energy, explained that UK Energy
Policy was to remove distortion in the market and leave
policy to market forces; the policy was in effect to have
no policy.   That is a conclusion forced upon an admin-
istration whose members do not have the scientific and
technical background to enable them to make long-term

planning decisions in a technological area. In the short
term the market in energy works, particularly when the UK
is fortunate in having indigenous coal, oil, gas and
nuclear power readily available, but in the long term,
where decisions must be made now about the fusion and fast
reactor programme, the coal-oil-renewables-conservation
mix, environmental control, the Severn Barrage and other
long lead time options, to leave the outcome to the market
could prove a very doubtful policy.

ENERGY PLANNING

The privatisation of the energy supply industries
with the promise of improved efficiency as a result of the
influence of the market-place can bring short-term
benefits but is a disincentive to long-term strategic
energy planning. There is no depletion strategy for North
Sea oil and gas, and with the suggestion of more, smaller
(300 MW) gas fired power stations following privatisation
the gas reserves will be rapidly depleted. If the 15,000
MW of additional electrical supply postulated by the CEGB
at the Hinkley Enquiry (Nov. 1988) as being necessary by
the year 2000 is to be gas-fired, once built and running
it would deplete in 20 short years the proven and probable
gas reserves (1325 x $10^9 m^3$ at July 1988) by 40 per cent.

The market-place also says nothing about
environmental protection, which can only be an added
expense and will only be implemented by setting legal
limits on emission. Similarly energy conservation is
anathema to energy supply companies trying to maximise
profits and return on capital. No wonder the various
energy efficiency campaigns have had only marginal
success. If the government were serious about energy
efficiency it would provide encouragement by fiscal means
at company and personal level allowing money spent on
improving the efficiency of energy usage as a tax
deductible expense.

The biggest problems facing energy planners is what
to do about the 'greenhouse effect' and to a lesser extent
'acid rain'. Both can only be mitigated by burning less
fossil fuel, particularly coal, although in the case of
acid rain some amelioration can be achieved by fitting
expensive 'scrubbing' equipment. If more nuclear stations
are to be built for electricity generation the world will
have to move to the Fast Reactor, which uses uranium fuel
60 times more efficiently than today's thermal stations
(the UK government removed support for this project in

July 1988).   The long-term fall-back position is the
fusion reactor.

      Nuclear power currently provides 20 per cent of
world electricity, if this were raised to 50 per cent it
would reduce global warming by 15 per cent.   An increased
use of renewable energy resources, particularly hydro,
tidal and wind power, is possible, but the World Energy
Conference has pronounced its disappointment over the
progress of renewable resources and believes 10 per cent
of world energy demand the best that can be achieved by
renewables well into the next century.   The latest (1988)
UK analysis provided by the Energy Technology Support Unit
in Energy Paper 55 confirms this view.   The maximum
contribution to electricity supply that the renewables,
wind, tidal, geothermal, wave and hydropower, can make in
2025 are estimated as 70 TWh/year, which is 35 million
tons of coal equivalent, about 10 per cent of current
primary energy demand.   It could be very much less than
that.

      CONCLUSION

      A coordinated attack on the world's energy problems
is essential if environmental damage, even disaster, is to
be   contained.   Attitudes   to   nuclear   power,   energy
efficiency and the profligate combustion of fossil fuels
will have to be rethought;   population growth contained;
and   even   some   countries'   expectations   modified.   A
strategy based on market forces will solve none of these
problems;   the industrialised world will have to help the
developing   world,   eager   to   increase   lifestyle   and
therefore energy consumption growth, by technology and
financial transfer.   Even with such a coordinated attack
on   the   problem   it   will   be   a   close-run   thing.

Acid Deposition

Ø. Hov
AVDELING FOR METEOROLOGI, UNIVERSITET I BERGEN, ALLEGT. 70, N-5007
BERGEN, NORWAY

1 INTRODUCTION

Acid deposition is defined as the total hydrogen ion
loading on a given area over a given period of time. The
total hydrogen ion deposited on a given area results from
acid rain, snow, fog, aerosols and gases such as nitric
acid and sulphur dioxide. Sulphur dioxide and nitrogen
oxides are the main precursors of acid deposition.
Oxidants such as ozone, hydrogen peroxide and organic free
radicals are required for the production of sulphuric and
nitric acid, giving rise to acid deposition. A schematic
picture of the acid deposition precursors and products is
shown in Figure 1[1].

Man-made emissions of atmospheric pollutants add
significant amounts of acid compounds to the environment.
Indications are strong that the sulphur dioxide (SO_2)
emissions increased by about a factor of two between 1950
and 1970 in most of Europe and the same general trend also
holds for nitrogen oxides (NOx). The emissions of SO_2 have
generally levelled off since 1970, while the NOx emissions
have continued to increase with the number of motor
vehicles. The deposition of sulphate and nitrate with rain
(wet deposition) accounts for 60-70% and 30-40% of the
total deposition of sulphate and nitrate in Europe,
respectively. The relative contribution of nitrate has
increased during the past 30 years. The acidifying
potential of wet deposition is also influenced by ammonia.
Ammonia neutralizes acids in rainwater, but in the soil
ammonia is eventually converted into nitrate by nitrifying
bacteria.

The rate of wet deposition of sulphur ranges from less than 3 kg S ha^{-1}a^{-1} in the Scandinavian countries, to more than 30 kg S ha^{-1}a^{-1} in rural areas in central Europe. The corresponding figures for nitrate are less than 0.1 kg N ha^{-1}a^{-1} in north Scandinavia, compared with more than 20 kg N ha^{-1}a^{-1} in central Europe.

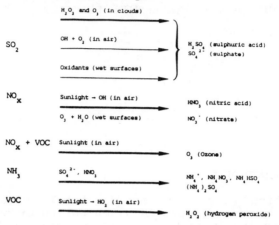

Figure 1 Acid deposition precursors and products

Table 1 Annual uptake of nitrogen and sulphur for high yielding crops, average fertilization rates of nitrogen, and wet deposition of sulphur and nitrogen during the growing season in the United States[1]

	kg ha^{-1}					
	NITROGEN (N)			NO_3 - N	SULPHUR (S)	SO_4-S
CROP	% FERTILIZED	UPTAKE	APPLICATION RATE	WET DEPOSITION	UPTAKE	WET DEPOSITION
Corn	96	298	153	1.5	37	3.8
Wheat	72	153	67	1.3	23	3.0
Soybean	20	363	20	1.6	28	4.2
Cotton	68	201	91	1.3	34	2.7

Nitrogen is an important plant nutrient, and the sources of nitrogen to crops include soil release, lightning, fertilizer application and wet and dry deposition originating from NOx emitted from combustion. Sulphur is also a nutrient essential for plant growth, but is needed at much lower levels than nitrogen. In Table 1 is shown the annual uptake of nitrogen and sulphur for high yielding crops, average fertilization rates of nitrogen, and wet desposition of sulphur and nitrogen during the growing season in the United States[1]. The numbers are also relevant for European conditions.

2 TRENDS IN ACID DEPOSITION

Ice cores from south Greenland have been analysed with respect to trace impurities such as chloride, nitrate and sulphate. Mayewski et al.[2] analysed an ice core drilled about 40 km away from a previous core site (Dye 3) in south Greenland. The core covered the period 1869 to 1984. The data showed that excess (nonseasalt) sulphate had increased threefold since approximately 1900, and the nitrate concentration had doubled since 1955. They attributed the increases to the deposition of those species from air masses carrying North American and European anthropogenic emissions. Neftel et al.[3] reported similar findings.

Chemical analysis of ice core samples from Dye 3, Greenland showed that nitrate fluxes had risen to about 30% above natural levels by the 1950s and were approximately proportional to anthropogenic input. There was a peak in the deposition of nitrate and sulphate in the late winter and early spring, coinciding with the Artic haze maximum[4].

The composition of precipitation (sulphate, nitrate, pH, other ions) and of air (SO_2, sulphate) has been measured in the EMEP network throughout Europe since 1978. An analysis of the measurements for 1978-1980 as compared to 1983-1985 for EMEP sites in Northwest Europe, indicated that there had been a significant decrease both in SO_2, sulphate aerosol, and sulphate and nitrate in precipitation in air which had passed over the British Isles[5]. The decrease in SO_2 and sulphate aerosol and sulphate in precipitation over West Europe during the last decade, is supported also by other measurements. In Figure 2 is shown the decrease in annual mean SO_2 concentration at 13 rural stations in the Federal Republic of Germany (FRG) in 1988 compared to the 1980-1987 average[6]. The significant decrease is attributed to the combined effects of an emission reduction in the FRG and its western neighbours, and two mild winters and a cool summer in between with predominantly western airflow.

In a comprehensive study where model calculations were kept together with EMEP measurements, Mylona[7] found a 18.9% decrease in measured annual SO_2 in the EMEP net-

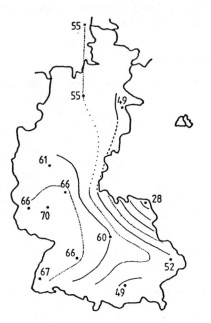

Figure 2 Decrease in annual mean SO_2 concentrations in
          1988 based on 13 rural stations in the FRG,
          compared to 1980-87, in per cent[6]

work in 1986 compared to 1980, while the calculated annual
averages decreased by 17.9%, in parts in response to a
16.4% reduction in the SO_2 emissions in Europe from 1980
to 1986 (26058 ktS in 1980, 21795 ktS in 1986). For
sulphate in air the measured decrease in annual concentration in 1986 was 15.5% compared to 1980, the
calculated decrease was 15.3%. Mylona[7] has made a very
interesting study of the influence of meteorology on the
calculated concentration changes. In Figure 3 is shown the
results for 42 rural EMEP sites for the annual average
concentration for 1979-1986 of SO_2 using actual emissions
and 1980 emissions and using actual emissions and 1980
meteorology. It can be seen that the 1980 meteorology gave
rise to a lower annual mean SO_2 concentration than most of
the subsequent years' meteorology keeping the emissions
constant. For the calculations where the emissions were
kept at 1980-level, while the meteorology corresponded to
the actual time, there is approx. a 20% difference between
the highest and lowest annual mean SO_2 concentration
during the 8 year period 1979-1986. This difference is
comparable to the emission reduction during the same

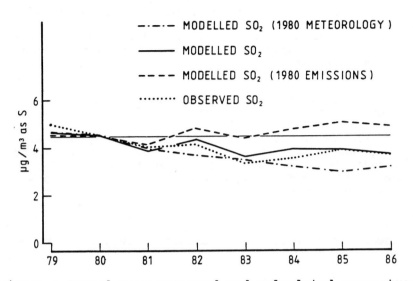

Figure 3 Annual mean measured and calculated concentrations of SO_2 over Europe, also showing calculated
concentrations assuming constant 1980 meteorological conditions throughout all the years, and
assuming 1980 emissions throughout. The horizontal line shows the calculated 1980 concentration
(42 measuring sites used)[7]

period, and this demonstrates that the change in meteorology even on an annual basis, can obscure the effect of
emission reductions on measured concentrations.

Precipitation data for EMEP for the period 1980-84
indicate that there is an increase in the annual mean
nitrate to non-marine sulphate equivalent ratio[8]. An
analysis of the nitrate content of monthly precipitation
samples collected in the European Air Chemistry Network
(EACN) during the period 1955 to 1979 showed that all
stations but one had an increase in concentration during
the 25 year period. The annual increase in nitrate
concentration between the late 1950s and early 1970s was
in the range 3-6% at 75% of the stations[9].

Measurements of nitrate aerosol at Harwell in the UK
from 1954-1982[10] showed an upward trend of approx. 3%/a for
both summer and winter for the whole period, with a
particularly strong increase up to about 1975 and more of
a levelling off thereafter. Seasonally averaged PAN

concentrations[11] at Delft, the Netherlands increased by
about a factor of 3 from 1973 to 1985, reflecting increases in both NOx and volatile organic compounds (VOC).
Indirect evidence that nitrate in precipitation may have
increased can be found in the results of "The 1000 lakes
study" in southern Norway, where the nitrate level in 305
lakes in 1986 was approximately twice the levels in the
same lakes in 1974-75[12].

3 MODEL CALCULATIONS OF TRENDS IN SO_2 AND NOx CONVERSION RATES

On a long-term basis and averaged over a large area, e.g.
for the annual mean over Europe, there is a linear
relationship between the change in emissions of SO_2 and
the deposition of S-compounds. This is probably nearly the
case also for the relationship between a change in NOx
emissions and the deposition of compounds derived from
NOx.

The conversion of SO_2 to sulphate mainly takes place
through

$$OH + SO_2 \rightarrow HSO_3 \qquad\qquad (i)$$

$$HSO_3 + O_2 \rightarrow HO_2 + SO_3 \qquad\qquad (ii)$$

where SO_3 reacts rapidly with water vapour

$$SO_3 + H_2O \rightarrow H_2SO_4 \qquad\qquad (iii)$$

When species like H_2O_2 and O_3 are dissolved in cloud water,
they readily oxidize dissolved SO_2. The H_2O_2-mechanism is
rather insensitive to the acidity of the water, while the
ozone-mechanism is very slow at low pH. Gaseous hydrogen
peroxide is a major oxidant leading to sulphuric acid
generation in cloud water. Emissions of NO are converted
to NO_2 through the very fast reaction

$$NO + O_3 \rightarrow NO_2 + O_2 \qquad\qquad (iv)$$

Conversion to nitric acid HNO_3 takes place through

$$OH + NO_2 \rightarrow HNO_3 \qquad \text{(v)}$$

which is about 10 times faster than the rate constant for the SO_2-reaction. The OH-radical vanishes at night. Measurements both in the U.S.[13] and in Europe[14] show that there is a strong seasonal dependence in the sulphate deposition suggesting a dominating photochemical influence on the gas and liquid phase chemistry. For the deposition of nitrate, there is no strong variation with season. Both OH and H_2O_2 vary strongly with season, supporting that a major part of the oxidation of SO_2 takes place through reaction with OH in the gas phase and H_2O_2 in the liquid phase.

The formation of nitrate in the winter and in the dark mainly takes place through

$$O_3 + NO_2 \rightarrow NO_3 + O_2 \qquad \text{(vi)}$$

$$NO_3 + NO_2 \rightleftharpoons N_2O_5 \qquad \text{(vii)}$$

$$N_2O_5 + H_2O \text{ (cloudwater)} \rightarrow 2H^+ + 2NO_3^- \qquad \text{(viii)}$$

Nitrate dissolved in water droplets is in equilibrium with nitric acid (HNO_3). Nitric acid is also in equilibrium with ammonium nitrate through

$$HNO_{3(g)} + NH_{3(g)} \rightarrow NH_4NO_{3(s)} \qquad \text{(ix)}$$

where g denotes gas and s solid. Ammonium nitrate dissolves into ammonium and nitrate in water. The equilibrium constant is dependent on temperature and relative humidity[15]. NO_2 is also converted into peroxyacetylnitrate through

$$NO_2 + CH_3COO_2 \underset{\Delta}{\rightleftharpoons} PAN \qquad \text{(x)}$$

where Δ is strongly temperature-dependent (slow at low temperatures).

In photochemical episodes, both OH, CH_3COO_2, H_2O_2 and O_3 increase in concentration, which means that the conversion of NOx and SO_2 to nitrate, PAN and sulphate is faster than usual. The dry deposition velocity of nitrate and sulphate aerosols is smaller than for SO_2 or NO_2. There is usually very little precipitation in photochemical episodes, favouring a long atmospheric lifetime of S- and NOx-species. On the other hand winds are usually low and vertical mixing poor in stagnant, anticyclonic weather conditions, which means that the extent of long-range transport or mixing with free tropospheric air do not increase as much as the lifetime increases.

Acid deposition is of concern in episodes, but environmental damage is mainly linked to the integrated acid deposition over a long time period. This has become very evident in South Norway, where the doubling of the nitrate content of freshwater lakes has taken place between 1974-75 and 1986, probably without a significant change in the nitrate deposition, indicating a saturation effect[12]. An attempt has been made[16] to show how slow changes in the emission of NOx and CH_4 could influence the formation of sulphate and nitrate in the global atmosphere. A global, 2-dimensional meridional model which is described elsewhere[17,18], was applied to calculate the changes in the global distribution of OH, H_2O_2, O_3 and NO_2 as the emissions of NOx and CH_4 were changed separately. Long-term calculations were carried out with 1980 as the base year for two different cases: An increase in the anthropogenic NOx-emissions by 3%/a, that is 2.1%/a in the total NOx-emissions, and 1.5%/a increase in the methane concentrations. The calculations were speeded up by computing explicitly only the first year, the base year (1980) and the last year. For methane a 60 yr period was covered with 1950 as the first and 2010 as the last year. The response time to changes in the emissions was so fast that sufficient convergence was achieved already after one year of model simulation. For methane it was chosen to change the concentration by approximately the amount observed to be the annual increase globally, rather than to increase the emission rate since it would require a calculation covering several decades for methane to reach steady state.

In Figure 4 is shown the meridional distribution at the end of May of O_3, OH and H_2O_2 with height in 1965 and 1995 in a 30 years' calculation with an annual increase in the anthropogenic NOx-emissions of 3%. In the southern hemisphere there was not much change in the concentrations, while in the lower part of the northern hemisphere up to 0.9%/a increase in ozone was calculated, 1.4%/a in OH, 2.1%/a in NOx while there was a drop in H_2O_2 of 0.5%/a or less.

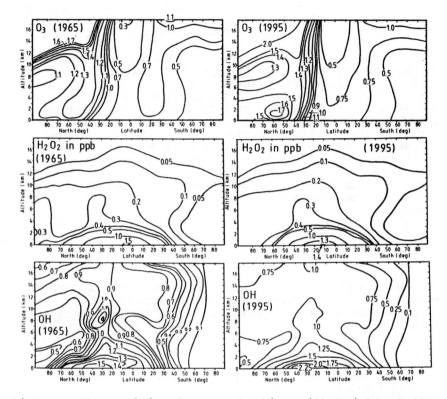

<u>Figure 4</u> The meridional concentration with height at the
end of May for O_3 (in 10^{12} molecu-
les/cm³), OH (in 10^6 molecules/cm³) and H_2O_2 (in
ppb) in 1965 (left column) amd 1995 (right
column) for a calculation where the global
anthropogenic NOx-emissions increased by 3%/a[16]

The increase in NOx was largely as NO_2. This means that
the formation of nitric acid through reaction (v) increased more than the increase in total NOx-emission in
the lower troposphere at mid northerly latitudes. The rate
of formation of nitric acid increased by nearly (2.1 +
1.4)%/a or 3.5%/a there. The increase in OH that was
calculated for increasing NOx-emissions would also enhance
gas-phase sulphate production through reactions (i)
followed by (ii) and (iii). The wet-phase conversion of
dissolved SO_2 to sulphate would proceed at a slower rate,

however, since it is mainly controlled by H_2O_2 which was calculated to decline as NOx goes up. This is due to the suppression of HO_2-radicals with increasing NO. At high pH wet-phase sulphate formation could be enhanced since dissolved ozone is active in the oxidation.

In Figure 5 is shown the meridional distribution at the end of May of O_3, OH and H_2O_2 with height in 1950 and 2010 in a calculation where the concentration of CH_4 increased by 1.5%/a over 60 years. Changes in O_3, OH and H_2O_2 were calculated to occur throughout the troposphere and in both hemispheres. In the lower part of the troposphere in the northern hemisphere O_3 was calculated to increase by about 0.4%/a, OH to decrease by 0.5%/a and H_2O_2 to increase by about 0.5%/a. The decrease in OH arose from the increase in its loss through reaction with CH_4. The concentration of hydroxyl decreased less than CH_4 went up, however, which means that the flux of material through the reaction

$$CH_4 + OH \rightarrow CH_3 + H_2O \qquad\qquad (xi)$$

increased and led to the formation of more O_3 and HO_2-radicals in the decomposition reactions of CH_4, and consequently also H_2O_2 went up since the production is determined by

$$HO_2 + HO_2 \rightarrow H_2O_2 + O_2 \qquad\qquad (xii)$$

The calculations showed that to increase CH_4 only, leads to a decrease in the conversion of NO_2 and SO_2 to nitrate and sulphate through the OH-reactions. The formation of nitrate in the dark would increase due to the O_3 increase, and the liquid phase conversion of SO_2 would likewise become more efficient. Reaction (viii) was not included in the model calculations, but the results from the model runs with changes in NOx-emissions and CH_4-concentrations can be used to get an indication about the change in the nitrate formation during the winter and during the night, provided that a process like reaction (viii) is the main pathway for nitrate formation. The rate of formation of the NO_3-radical at night is controlled by reaction (vi). The formation rate of the NO_3-radical increases as the sum of the increase in NO_2 and O_3. The concentration of N_2O_5 is given by the equilibrium (vii) and is shifted to the right which means that $N_2O_5 >> NO_3$ at

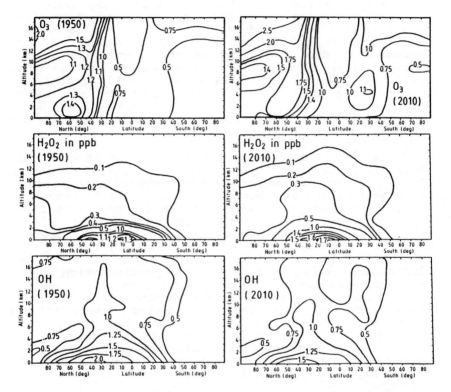

<u>Figure 5</u>   The meridional concentration with height at the
end of May for O_3 (in 10^{12} molecules/cm^3), OH (in
10^6 molecules/cm^3) and H_2O_2 (in ppb) in 1950 (left
column) and 2010 (right column) for a calculation
where the tropospheric concentration of CH_4 was
increased by 1.5%/a[16]

night at atmospheric temperatures. This means that the
formation rate of nitrate through reaction (viii) at night
in the lower mid latitude northern troposphere increases
as the sum of the increases in NO_2 and O_3.

It can be concluded that to increase the global NOx-
emission enhances both the photochemical conversion of NO_2

to HNO_3 through reaction with OH, and even more enhances the efficiency of the dark reaction where N_2O_5 reacts with water droplets. The sulphate formation is increased since OH goes up, while the liquid phase conversion becomes less efficient due to a suppression of H_2O_2 as NOx increases. A picture of the combined effect of increasing the anthropogenic NOx-emissions by 3%/a and the CH_4-concentrations by 1.5%/a can be gained by just adding the calculated annual changes in the concentrations together. At mid latitudes in the lower part of the troposphere in the northern hemisphere, O_3 would increase at a rate of 1.3%/a, OH increase by 0.9%/a, NOx would increase by 2.1%/a while H_2O_2 would remain virtually unchanged. Measurements of H_2O_2 in ice samples drilled on North Central Greenland seem to support that H_2O_2 in the air has not changed much in this century[3].

The implication of these numbers is that at mid latitudes in the lower troposphere, the rate of formation of HNO_3 through the OH-route would increase somewhat more than the increase in NOx-emission. The rate of formation of sulphate through the OH-route would increase by about 0.9%/a plus the rate of increase in the SO_2 concentration, while the liquid phase conversion would increase approximately at the same rate as the increase in the SO_2 concentration. The winter mean concentration of the sum of nitrate and nitric acid in air and precipitation should have increased more than the summer values at mid latitudes in the northern hemisphere during years with increasing NOx-emissions. The findings also suggest that the fraction that nitric acid + nitrate makes up of lower tropospheric NOy (NOx + PAN + HNO_3 + nitrate), has increased more than the fraction that sulphate makes up of lower tropospheric SOx (=SO_2 + sulphate) during times with increasing SO_2 and NOx emissions.

ACKNOWLEDGEMENT

The global model calculations were done in collaboration with Professor Ivar Isaksen, Institute of Geophysics, University of Oslo. The meteorological variability of the annual average SO_2 concentration is based on the work of Sofia Mylona at MSC-W, The Norwegian Meteorological Institute, Oslo. The work reported here is partly sponsored by the Nordic Council of Ministers.

REFERENCES

1. NAPAP, 'The National Acid Precipitation Assessment Program', Washington, D.C., 1987.
2. P.A. Mayewski, W.B. Lyons, M.J. Spencer, M. Twicker, W. Dansgaard, B. Koci, C.I. Davidson and R.E. Honrath, Science, 1986, 232, 975.
3. A. Neftel, P. Jacob and D. Klockow, Nature, 1984, 311, 43.
4. R.C. Finkel, C.C. Langway Jr. and H.B. Clausen, _ J.Geophys. Ser., 1986, 91, 9849.
5. K. Nodop and H.W. Georgii, in 'Proceedings of the Fifth European Symposium on the Physico-Chemical Behaviour of Atmospheric Pollutants', Reidel, Dordrecht, 1990.
6. S. Beilke, 'Summary by the Chairman' in 'Proceedings of the Fifth European Symposium on the Physico-Chemical Behaviour of Atmospheric pollutants' Reidel, Dordrecht, 1990.
7. S.N. Mylona, EMEP MSC-W Report 1/89, The Norwegian Meteorological Institute, Oslo, 1989.
8. K. Nodop, in 'Proceedings of the Fourth European Symposium on Physico-Chemical Behaviour of Atmospheric Pollutants', Reidel, Dordrecht, 1986.
9. R. Söderlund, L. Granat and H. Rodhe, Report CM-69, University of Stockholm, Dep. of Meteorology, 1985.
10. PORG, 'Oxides of Nitrogen in the United Kingdom', A Second Report of the UK Photochemical Oxidants Review Group, 1989.
11. R. Guicherit, in 'Tropospheric Ozone', I.S.A. Isaksen (ed.), Reidel, Dordrecht, 1988, 48.
12. B. Bergmann-Paulsen and B. Kvæven, SFT Report 283/87, SFT, Oslo, 1987.
13. J.G. Calvert, A. Lazrus, G.L. Kok, B.G. Heikes, J.G. Walega, J. Lind and C.A.Cantrell, Nature, 1985, 317, 27.
14. J. Schaug, J.E. Hanssen, K. Nodop, B. Ottar and J.N. Pacyna, EMEP-CCC Report 3/87, NILU, Lillestrøm, 1987.
15. A.W. Stelson and J.H. Seinfeld, Atmos. Environm., 1982, 16, 983.
16. Ø. Hov and I.S.A. Isaksen, in 'Proceedings of the Fourth European Symposium on the Physico-Chemical Behaviour of Atmospheric Pollutants', Reidel, Dordrecht, 1986.
17. I.S.A. Isaksen, Ø. Hov, S.A. Penkett and A. Semb, J. Atmosph. Chem., 1985, 3, 3.
18. I.S.A. Isaksen and Ø. Hov, Tellus, 1987, 39B, 271.

Tropospheric Effects

R. G. Derwent
AEA TECHNOLOGY, HARWELL LABORATORY, EMSCD, BUILDING 364,
DIDCOT, OXFORDSHIRE OX11 0RA, UK

1 INTRODUCTION

The conversion of most fuels into useful energy for heating homes, driving motor vehicles, powering industry and generating electricity leads inevitably to pollution of the environment. The last few decades have seen an intensifying interest in global scale pollution problems and the role played by fuel use and energy conversion processes. Pollution abatement, safety procedures and environmental concerns have significantly influenced the economics of competing energy conversion processes. In this review, attention is given to global scale environmental problems to see how their consideration may influence the competitive position of the main energy carriers over the coming decades.

The global scale environmental problems considered here stem from the growing concern about changes in the atmospheric composition of the troposphere. There is a substantial body of evidence that concentrations of a range of trace gases are increasing due to man's activities. Some of these trace gases including carbon dioxide, methane, chlorofluorocarbons, nitrous oxide and ozone are radiatively active gases. Their accumulation in the troposphere may increase global warming at the earth's surface through the "Greenhouse Effect".

Attempts to minimise the extent of any future global warming depend on an understanding of the relative importance of each trace gas and the cost and availability of technical measures for their control. It is crucially important to understand quantitatively the role played by energy conversion processes as a global scale source of each trace gas and the relative efficiency for each trace gas in contributing to global warming.

The first part of this review addresses the concept of an

index of global warming efficiency for each trace gas on an
emissions basis. The future commitment to global warming
following the emission of a pollutant into the troposphere depends
on three factors, as follows:

o   the different infrared absorption properties of each trace gas,

o   the different atmospheric lifetimes, which govern the long
     term behaviour of each trace gas,

o   the influence of each trace gas on the concentration of other
     greenhouse gases.

These factors are reviewed for a wide range of trace gases and
their influence on the future warming commitment following unit
release has been determined. In the second part, the atmospheric
life cycles of these trace gases are examined to uncover the role
and influence of energy conversion processes in their emissions.

2 GLOBAL WARMING POTENTIALS OF GREENHOUSE GASES

The approach implied by the concept of a Global Warming
Potential based on emissions considers two main classifications of
radiatively active gases:

o   those trace gases which exert a <u>direct</u> radiative impact,

o   those trace gases which react chemically in the troposphere to
     produce carbon dioxide <u>or</u> modify the tropospheric distribution
     of ozone <u>or</u> modify the stratospheric distribution of water
     vapour and hence exert an <u>indirect</u> radiative impact.

Trace Gases with a Direct Radiative Impact

The situation is imagined where 1 kg of a trace gas is
released into the atmosphere and is instantaneously and completely
mixed throughout the atmosphere. The trace gas is then subject to
atmospheric removal and its concentration starts to decline. Whilst
present in the atmosphere, the radiatively active gas absorbs and
reemits long wavelength radiation emitted by the earth's surface.
At any point in time after release, the trace gas will produce a
surface warming measured in $^{\circ}C$ which when integrated over a fixed
time horizon, implies a commitment to global warming measured in $^{\circ}C$
x years. The aim is to tabulate global warming commitments for the
release of 1 kg of a range of radiatively active gases.

The concentration-time behaviour of the trace gas in the
scenario outlined in the paragraph above, can be estimated by
solving the differential equation for a well-mixed box representing
the atmosphere. Then

Table 1   Estimates of ΔC, a and ΔW values for five trace gases which exhibit a direct radiative impact.

Trace Gas	τ Years	ΔC, ppm years			Warming Contribution, c $^\circ$C ppm^{-1}	ΔW, $^\circ$C years		
		Time horizon, years				Time horizon, years		
		20	100	500		20	100	500
carbon dioxide	und[a]	1.9×10^{-12}	6.9×10^{-12}	2.0×10^{-11}	0.0049	9.3×10^{-15}	3.4×10^{-14}	1.0×10^{-13}
methane	10[b]	3.1×10^{-12}	3.5×10^{-12}	3.5×10^{-12}	0.13	4.0×10^{-13}	4.6×10^{-13}	4.6×10^{-13}
nitrous oxide	150	2.4×10^{-12}	9.4×10^{-12}	1.9×10^{-11}	0.81	1.9×10^{-12}	7.6×10^{-12}	1.5×10^{-11}
CFC 11	58	6.9×10^{-13}	2.0×10^{-12}	2.4×10^{-12}	54	3.6×10^{-11}	1.1×10^{-10}	1.3×10^{-10}
CFC 12	130	8.7×10^{-13}	3.3×10^{-12}	5.9×10^{-12}	67	5.8×10^{-11}	2.2×10^{-10}	4.0×10^{-10}

Notes

a. the values for CO_2 have been obtained from the integration of the decay of an incremental 1 kg of CO_2 added to the box-diffusion model of the coupled atmosphere-biosphere-ocean system[2].

b. reference 7 quote a lifetime for methane in the range 8.1-11.8 years with a best estimate of 9.6 years.

c. based on references 8 and 9 for small perturbations on current trace gas concentrations.

$$\frac{dc}{dt} = - \frac{c}{\tau} \tag{1}$$

where c is the average volume mixing ratio of the trace gas, t is
    time,
and τ is the lifetime of the trace gas, so that:

$$c = Co \; exp^{-\frac{t}{\tau}} \tag{2}$$

where Co is the initial concentration at time t = 0, 1st January.

Integrating equation (2) over the time horizon, t_h, then the
total commitment to elevated concentrations of that trace gas, ΔC,
becomes:

$$\Delta C = \int_{o}^{t_h} C_o \; exp \left[- \frac{t}{\tau} \right] dt \tag{3}$$

For 1 kg release of a trace gas of molecular weight, MW, then,

$$\Delta C = \frac{\tau}{MW} \; \frac{MW_{air}}{5.136 \times 10^{18}} \; x \; 10^6 \; x \; \{1 - exp \; (-t_h/\tau)\} \; ppm \; years \tag{4}$$

where MW_{air} is the molecular weight of the atmosphere,
and 5.136×10^{18} is the atmospheric mass[1] in kg.

Table 1 gathers together the results of applying equation (4) to
the calculation of ΔC values for a number of trace gases of
interest. In all cases, the atmospheric lifetimes of the trace
gases have not been accurately determined. A range of estimates is
available from the literature and the value tabulated is some form
of 'best estimate' or 'central value'.

For most of the trace gases of interest in Table 2, this
simple approach represents a reasonable simplification of a range
of complex atmospheric processes. The assumption of complete and
instantaneous mixing would be a problem, for example, for any trace
gas with an atmospheric lifetime approaching one year because of
the neglect of spatial gradients. The assumption of first order
decay implies that removal processes are independent of trace gas
concentration and that lifetimes remain constant throughout the
integration period. This assumption may not be valid for all of
the species in Table 2. For those trace gases which are removed
by tropospheric OH oxidation, this assumption implies that hydroxyl
radical concentrations remain unchanged. This will not necessarily

Table 2 Estimates of Δ(Global Ozone) and ΔW for the instantaneous release of 1 kg of a trace gas integrated over 20, 100 and 500 years time horizons through its impact on tropospheric ozone production.

Trace Gas	Δ(Global Ozone Concentration),		ΔW, °C years		
	ppm[a]	ppm years	20	100	500 years
oxides of nitrogen[c]	1.9×10^{-13}	1.9×10^{-13}	1.5×10^{-12}	1.5×10^{-12}	1.5×10^{-12}
carbon monoxide	6.4×10^{-15}	6.4×10^{-15}	5.1×10^{-14}	5.1×10^{-14}	5.1×10^{-14}
methane	3.5×10^{-15}	3.5×10^{-14}	2.4×10^{-13}	2.8×10^{-13}	2.8×10^{-13}
non-methane hydrocarbons	3.6×10^{-14}	3.6×10^{-14}	2.9×10^{-13}	2.9×10^{-13}	2.9×10^{-13}

Notes:

a. calculated using the Harwell Global two-dimensional model.

b. global warming estimated by assuming 12.5% increase in the mean tropospheric ozone concentration about a 1980 value of 61 ppb gives a 0.06°C warming[8], ppm^{-1}.

c. 1 kg of NO_x expressed as NO_2.

e. these ΔW values are highly model dependent because of the importance of non-linearity in the relationship between tropospheric ozone production and trace gas emissions and because of the difficulty of folding in together the latitude, altitude and seasonal changes in the tropospheric ozone distribution with their radiative impact.

be the case in some future situations, depending on methane, carbon monoxide and nitrogen oxide emissions.

Carbon dioxide is difficult to handle in such a simple approach because of the complex nature of its exchange processes with the biosphere and with the oceans. The approach adopted in this study uses the box-diffusion model approach to carbon cycle modelling[2]. The response of the atmosphere-biosphere-ocean system to a pulse of carbon dioxide has been studied in the box-diffusion model and its decay has been found to be markedly non-exponential. The integrals in equation (3) for carbon dioxide have been found directly from the box-diffusion model and have not relied on the assumption of a simple lifetime.

To complete the calculation of global warming commitments, estimates are required of the warming contribution on a molar basis. These estimates are given in Table 2, which also gathers together the commitments to elevated concentrations and warming contributions to produce estimates of commitment to global warming for the instantaneous release of 1 kg of trace gas, integrated over a range of time horizons.

Inherent in this approach is the assumption of a relationship such as:

$$\Delta W = a \, \Delta C \qquad (5)$$

where ΔC is the commitment to elevated concentrations in ppm years
a is the warming contribution in $^{\circ}C$ per ppm
and ΔW is the global warming commitment in $^{\circ}C$ years, following the release of 1 kg of the trace gas.

Equation (5) remains a valid approximation only under conditions of infinitesimally small changes about present day trace gas concentration distributions, where there are no overlap of absorption bands by different trace gases, no chemical interactions between pollutants and that there are no changes in other radiative agents such as water vapour, clouds and aerosols. Extrapolation to realistic emissions may ignore important interactions between trace gases and may invalidate the linear assumptions inherent in this approach.

For some environmental impacts, it is important to evaluate the cumulative greenhouse warming over an extended period after the release of a trace gas into the atmosphere. For the evaluation of sea-level rise, the commitment to greenhouse warming over a 500 year or longer time horizon, t_h, may be appropriate. For the evaluation of short-term effects, a time horizon of 10 years or less could be taken. This consideration alone, dramatically changes the emphasis between the different trace gases, depending

on their persistence in the atmosphere. For this reason,
commitments to global warming have been evaluated over 20, 100 and
500 year time horizons. The global warming commitment is scaled
down by the factor of $1 - \exp(-t_h/\tau)$ from the infinite time horizon
value, where t_h is the time horizon and τ is the trace gas
lifetime. Choosing shorter time horizons scales down the
commitments from the longest- lived trace gases such as CO_2, N_2O
and the CFCs and scales up the contributions from CH_4, CO, NO_x and
the HFCs.

Trace Gases with Indirect Radiative Impacts

In this section, the situation is imagined where 1 kg of trace
gas is released into the atmosphere on the 1st of January and is
instantaneously and completely mixed throughout the atmosphere. The
trace gas is then subject to atmospheric removal and its average
concentration throughout the atmosphere starts to decline.
Whilst it is present in the atmosphere, the trace gas modifies the
tropospheric ozone balance. Since tropospheric ozone is a potent
radiatively active gas, the trace gas may exert an indirect
"Greenhouse Effect". The trace gas may degrade in the atmosphere to
produce carbon dioxide and thereby also exert an indirect
"Greenhouse Effect". Furthermore, methane may be degraded in the
stratosphere perturbing the water vapour distribution there,
leading to an additional indirect "Greenhouse Effect".

In the case of the indirect "Greenhouse Effect" due to
tropospheric ozone the basis approach has to be slightly modified,
as follows:

$$\Delta W = \Delta C \text{ (Global Ozone)} \times \text{Warming}$$
$$\text{Contribution per ppm of ozone} \qquad (6)$$

Implicit in this relationship is that the infinitesimal change in
global surface warming can be related to the change in global ozone
concentration, through a proportionality constant. In principle,
each trace gas will modify the tropospheric ozone distribution in
subtle, different ways. These differences have not been taken into
account. The only factor which has been taken into account is the broad
chemical differences between the various trace gases and their
impact on the global ozone concentration distribution.

Table 2 gathers together some of the results for a range of
trace gases which when released into the atmosphere stimulate
global tropospheric ozone production. These estimates have been
obtained by analysing the response of the Harwell Global
tropospheric two-dimensional model to small changes in the man-made
emissions of each trace gas and studying the impact on global
tropospheric ozone production. These responses have taken into
account the simultaneous impacts of these trace gases on the global

tropospheric OH distribution as well as on the ozone distribution and apply only to small perturbations around the present day global ozone distribution. Global tropospheric ozone changes are highly model dependent because of the inherent non-linearity in the relationship between the ozone production terms and the trace gas concentrations, particularly those of NO_x. These ozone changes are therefore highly uncertain and are likely to be only first order estimates and are included only for comparative purposes.

In each case a constant injection over one year was modelled and the impact estimated on the annual tropospheric ozone budget. The Δ (Global Ozone Concentration) was obtained from the ratio of the change in the tropospheric ozone budget to the change in the trace gas emission rate. All the results refer to the release of a pollutant at the surface of the midlatitudes of the northern hemisphere.

If, after the first year, the atmospheric release of the trace gas is terminated, the trace gas will decline in concentration and so will its contribution to the tropospheric ozone budget. For the majority of the trace gases in Table 2, their lifetimes are significantly shorter than one year so little ozone production survives into the second and subsequent years. Furthermore, ozone has a short lifetime in the troposphere, about 60 days, so that little of the ozone produced in the first year also survives into the second and subsequent years. In assessing the commitment to global warming over each time horizon therefore, no additional contribution has been taken into account for the second and subsequent years.

For methane, the situation is particularly different because of its significantly longer lifetime compared with the other gases in Table 2. In integrating over each time horizon, a lifetime factor of 10 years has been applied in exactly the same manner as in the determination of its direct "Greenhouse Effect".

A major area of uncertainty involves the evaluation of the climate impacts of tropospheric and stratospheric ozone concentration changes. Several studies have shown that the effects of ozone on climate will depend strongly on the changes in ozone distribution with altitude and latitude. Changes in the ozone distribution in the upper troposphere and lower stratosphere are the most effective in causing a surface temperature change, with increased ozone in this region leading to increased surface temperatures[3]. The indirect ΔW values in Table 2 are highly model dependent because of the uncertainties involved in folding together the spatial distribution of the ozone changes with their radiative impact.

Table 3   Estimates of ΔW for the instantaneous release of 1 kg of a
           trace gas which produces carbon dioxide in its atmospheric
           degradation, assessed over 20, 100 and 500 years time
           horizon.

         Trace Gas   ΔW, °C years

                              20   100   500 years

carbon monoxide 1.47×10^{-14} 5.34×10^{-14} 1.58×10^{-13}
  (fossil CO)

methane 2.56×10^{-14} 9.34×10^{-14} 2.77×10^{-13}
  (fossil CH_4)

non-methane hydrocarbons 2.93×10^{-14} 1.07×10^{-13} 3.16×10^{-13}
  (fossil)

For the carbon-containing trace gases derived from fossil
sources, atmospheric oxidation will ultimately lead to carbon
dioxide production and hence a contribution must be allowed for the
latter's radiative impact. These indirect commitments to global
warming over an infinite time horizon are detailed in Table 3 over
each time horizon.

The radiative impact of the changes in the distribution of
stratospheric water vapour produced by the oxidation of added
methane increases the direct "Greenhouse Effect" of methane[4] by
about 30%.

The Combined Direct and Indirect Radiative Impacts Following The Instantaneous Release of a Range of Trace Gases

In this section, the direct and indirect radiative impacts of
each trace gas are added together and compared in Table 4. The
values are tabulated in °C years in which the global warmings
following the instantaneous release of 1 kg of trace gas have been
integrated over an infinite time horizon. The global warming
values for each trace gas have then been expressed relative to
carbon dioxide on an equivalent mass emission basis. In this way,
it becomes apparent that all the trace gases in Table 4 have a
considerably greater radiative impact compared with carbon dioxide.
The relative impacts span three orders of magnitude between carbon
dioxide and CFC-12.

Table 4   The combined direct and indirect global warming
          commitments integrated over 20, 100 and 500 year time
          horizons following the instantaneous release of 1 kg of
          the trace gas.

Trace Gas		ΔW, $^{\circ}$C years		
		20	100	500 years
carbon dioxide		9.3×10^{-15}	3.4×10^{-14}	1.0×10^{-13}
methane	Direct	4.0×10^{-13}	4.6×10^{-13}	4.6×10^{-13}
	Indirect (O_3)[c]	2.4×10^{-13}	2.8×10^{-13}	2.8×10^{-13}
(fossil)	Indirect (CO_2)	2.6×10^{-14}	9.3×10^{-14}	2.8×10^{-13}
	Indirect (H_2O)	1.2×10^{-13}	1.4×10^{-13}	1.4×10^{-13}
nitrous oxide		1.9×10^{-12}	7.6×10^{-12}	1.5×10^{-11}
CFC 11		3.8×10^{-11}	1.1×10^{-10}	1.3×10^{-10}
CFC 12		5.8×10^{-11}	2.2×10^{-10}	4.0×10^{-10}
carbon monoxide[a,c] (fossil)		6.6×10^{-14}	1.0×10^{-13}	2.1×10^{-13}
(living)		5.1×10^{-14}	5.1×10^{-14}	5.1×10^{-14}
oxides of nitrogen[b,c]		1.5×10^{-12}	1.5×10^{-12}	1.5×10^{-12}
non-methane hydrocarbons[a,c]				
(fossil)		3.2×10^{-13}	4.0×10^{-13}	6.1×10^{-13}
(living)		2.9×10^{-13}	2.9×10^{-13}	2.9×10^{-13}

Notes:

a. this represents the sum of the direct and indirect global
    warmings from tropospheric ozone and carbon dioxide formation.

b. 1 kg of NO_x expressed as NO_2.

c. this entry includes a contribution from the radiative impact of
    tropospheric ozone which is particularly uncertain because
    estimates of changes in the tropospheric ozone distribution are
    themselves model dependent and because of the difficulties of
    folding together the spatial distribution of the ozone changes
    with their radiative impact.

In this study, the Global Warming Potential of the Emissions of a Greenhouse Gas is defined as the time integrated commitment to global warming from the instantaneous release of 1 kg of a trace gas expressed relative to that from 1 kg of carbon dioxide:

$$\text{GWP} = \frac{\int_0^{t_h} a_i \cdot C_i \, dt}{\int_0^{t_h} a_{CO_2} \cdot C_{CO_2} \, dt} = \frac{a_i \, \Delta C_i}{a_{CO_2} \, \Delta C_{CO_2}} = \frac{\Delta W_i}{\Delta W_{CO_2}}$$

where a_i is the instantaneous global warming due to a unit increase in concentration of trace gas, i,

C_i is the concentration of the trace gas remaining at time, t, after its release,

t_h is the integration time horizon over which the commitment is summed, with the corresponding values for carbon dioxide in the denominator.

Table 5 contains the GWP values for each trace gas for each time horizon.

This approach to the definition of Global Warming Potentials[5,6] studies the global warming following the instantaneous emission of a pulse of a particular trace and follow its decay in the atmosphere. It takes into account both the direct and indirect "Greenhouse Effects". It can be employed in the analysis of realistic emissions scenarios by splitting the latter up into a large number of instantaneous releases of different magnitudes over extended periods. Emission abatement scenarios can therefore be evaluated using this concept.

3 TRACE GAS EMISSIONS FROM THE ENERGY INDUSTRIES

Atmospheric research programmes have revealed that the tropospheric concentrations of a number of trace gases are increasing due to man's activities. For some trace gases, their atmospheric life cycles are not fully understood and it is more difficult to be certain about the role of man's activities. Emissions from the energy industries and energy conversion processes represent an important subset of source terms in these life cycles, along with agriculture, deforestation, cement manufacture, biomass burning, process industries and natural biospheric processes.

Table 5   The combined direct and indiect global warmings integrated
over a 20, 100 and 500 year time horizons following the
instantaneous release of 1 kg of the trace gas expressed
relative to carbon dioxide.

	Global Warming Potential Time Horizon, Years		
	20	100	500
carbon dioxide	1	1	1
methane[a] (fossil)	84	29	12
(living)	81	26	9
nitrous oxide	210	220	150
CFC-11	4000	3100	1300
CFC-12	6200	6400	4000
$CO^{d,e}$ (fossil)	7	3	2
(living)	6	2	0.5
$NO_x{}^d$	160	44	15
non-methane hydrocarbons[d,e] (fossil)	34	13	6
(living)	31	9	3

Notes:

a.   this represents the sum of the direct and indirect global
warmings from tropospheric ozone, stratospheric water vapour
and carbon dioxide formation.
b.   1 kg of NO_x expressed as NO_2.
c.   the equivalent carbon dioxide emissions are given to two
significant figures to facilitate comparison and not because of
intrinsic precision.
d.   this entry includes a contribution from the radiative impact of
tropospheric ozone which is particularly uncertain because
estimates of changes in the tropospheric ozone distribution are
themselves model dependent and because of the difficulties of
folding together the spatial distribution of the ozone changes
with their radiative impact.
e.   this represents the sum of the indirect global warmings from
tropospheric ozone and carbon dioxide formation.
f.   the Global Warming Potential value for a trace gas is expressed
as the mass emission of carbon dioxide in kg that produces the
same long term commitment to global warming as 1 kg of release
of the trace gas, integrated over 20, 100 and 500 year time
horizons.

Table 6 Annual global trace gas emissions from energy conversion
expressed as their global warming potentials relative to
carbon dioxide over different time horizons.

Trace Gas	Annual Emission Rate From Energy Conversion, 10^{12}g yr^{-1}	Global Warming Potential, $CO_2 = 100$			Comments
		Time horizon, years			
		20	100	500	
carbon dioxide[a]	19500	100	100	100	fossil fuel combustion
methane	20-50	9-22	3-7	1-3	natural gas losses
	12-40	5-17	2-6	1-2	coal mining
	30-60	13-26	4-9	1-3	solid waste burial
carbon monoxide	450	16	7	5	motor vehicles
nitrogen oxides (NO_x)[b]	20	16	5	2	coal burning
	10	8	2	1	oil burning
	7	6	2	1	gas burning
	19	16	4	1	transportation
non-methane hydrocarbons	90	3	1	1	sources are not well known

Notes:

a. expressed as CO_2.

b. expressed as $NO_x = NO_2$.

In Table 6 we have gathered together a crude inventory of the current global annual emissions of a range of trace gases due to the energy industries or energy conversion processes. The estimates are highly uncertain and serve only as rough estimates to guide understanding and discussion. For each time horizon; 20, 100 and 500 years, the global warming potentials of each trace gas have been expressed relative to carbon dioxide (CO_2 = 100). In all timescales the emissions of the other trace gases in addition to carbon dioxide are not negligible.

If a short term view of the commitment to global warming, say over the 20 year time horizon, then the contribution from other trace gases may approach and exceed that of carbon dioxide, itself. The methane sources from energy conversion processes and the energy industries have the potential to provide a significant short term contribution to global surface warming.

Over longer time horizons, the long atmospheric lifetime of carbon dioxide shows through as a major influence and the contributions from the other trace gases appear to be much smaller. Over a 500 year time horizon which may be appropriate for studying sea-level rise, for example, the other trace gases may add an additional contribution of 13-18 per cent in addition to that from CO_2 itself.

4 ACKNOWLEDGEMENTS

This work was supported as part of the United Kingdom Department of the Environment Air Pollution Programme under Contract No. PECD 7/10/172. The author wishes to acknowledge discussions with Colin Johnson and Adrian Hough of the Harwell Laboratory, Geoff Jenkins of the IPCC Centre, United Kingdom Meteorological Office and Henning Rodhe of the Department of Meteorology, University of Stockholm and Don Wuebbles of the Lawrence Livermore National Laboratory.

5 REFERENCES

1. K.E. Trenberth, J.R. Christy and J.G. Olson. Global atmospheric mass, surface pressure and water vapour variations. *J. Geophys. Res.*, 1987, *92*, 14815-14826.

2. U. Siegenthaler. A box diffusion model to study the carbon dioxide exchange in nature. *Tellus*, 1975, *27*, 168-192.

3.   A.A. Lacis, D.J. Wuebbles and J.A. Logan. Radiative forcing
     of climate by changes in the vertical distribution of ozone.
     J. Geophys. Res. (to be published), 1989.

4.   D.J. Wuebbles, K.E. Grant, P.S. Connell and J.E. Penner. The
     role of atmospheric chemistry in climate change. Journal of
     Air Pollution Control Association, 1989, 39, 22.

5.   D.A. Lashof and D.R. Ahuja. Relative global warming
     potentials of greenhouse gas emissions. Nature, in press,
     1989.

6.   H. Rodhe. Vaxt Luseffekten. Naturvardsverket Rapport 3647,
     Box 1302, 17125, Solna, Sweden, 1989.

7.   R.J. Cicerone and R.S. Oremland. Biogeochemical aspects of
     methane emissions. Global Biogeochemical Cycles, 2, 299-327,
     1988.

8.   V. Ramanathan, R.J. Cicerone, H.B. Singh and J.T. Kiehl.
     Trace gas trends and their potential role in climate change.
     J. Geophysical Research, 1985, 90, 5547-5566.

9.   I.M. Mintzner. A matter of degrees: The potential for
     controlling the greenhouse effect. World Resources Institute,
     Washington DC, USA, 1987.

The Current Status of Stratospheric Ozone

A. F. Tuck
NOAA/ERL, 325 BROADWAY, BOULDER, COLORADO 80303/3328, USA

1 INTRODUCTION

If the stratosphere is treated as a motionless pure oxygen photochemical reactor, Chapman[1], using the solar flux and the following reactions

$$O_2 + h\nu \ (\lambda < 245nm) \longrightarrow O + O$$
$$O + O_2 + M \longrightarrow O_3 + M$$
$$O + O_3 \longrightarrow O_2 + O_2$$
$$O_3 + h\nu \longrightarrow O + O_2$$

two gross discrepancies become apparent. The stratosphere is calculated to have an excess of ozone by a factor of two or more over the real atmosphere, Brewer and Wilson[2], and it is incorrectly distributed: the highest column abundances are at middle and high latitudes not, as calculated in this motionless system, in the tropics. These discrepancies signalled firstly the importance of chemical chain reactions involving other species, as first suggested by Hampson[3,4], who also suggested that water vapour from aircraft exhausts could affect the ozone balance; Crutzen[5] reinforced and expanded Hampson's suggestion of NO and NO_2, while Molina and Rowland[6] pointed out the importance of Cl and ClO:

$$X + O_3 \longrightarrow XO + O_2$$
$$XO + O \longrightarrow X + O_2$$

$$\text{nett: } O + O_3 \longrightarrow O_2 + O_2$$

where X = H, OH, NO, Cl and Br. The second discrepancy shows the role of fluid motions in transporting the ozone in the stratosphere. The fluid motions are driven by the radiative heat balance, which is in turn determined by the absorbing properties of ozone in the ultra violet and visible, and by the absorbing and emitting properties of CO_2, H_2O and O_3 in the infrared. The chemical composition is thus the result of an interactive, dynamic balance between photochemical, radiative and fluid mechanical processes.

The chain reactions may have lengths in the range 10^3 - 10^5, and this makes it feasible for trace species at mixing ratios of order 10^{-9} to effectively destroy ozone at mixing ratios of 10^{-5} or 10^{-6}. The chain carrying species OH & HO_2, NO & NO_2 are produced naturally by reaction of source molecules from the planetary surface with $O(^1D)$ Hampson[3,4]:

$$O(^1D) + H_2O \ \text{-->} \ OH + OH$$
$$O(^1D) + N_2O \ \text{-->} \ NO + NO$$

Some Cl & ClO molecules are produced naturally from methyl chloride:

$$OH + CH_3Cl \ \text{-->} \ H_2O + CH_2Cl \quad \text{-->} \ Cl \ \& \ products$$
$$CH_3Cl + hv \ \text{-->} \ Cl + CH_3$$

but the majority in the current stratosphere comes from CFC s, Molina & Rowland[6], for example:

$$CF_2Cl_2 + hv \ \text{-->} \ Cl + CF_2Cl$$
$$CFCl_3 + hv \ \text{-->} \ Cl + CFCl_2$$

The radiation needed to photodissociate CFC s is only available above 20km; since they are stable below this altitude, they have a long atmospheric lifetime, determined by the need to cycle the entire content below 20km (95% of the atmosphere) to altitudes above. All the chlorine on these molecules is believed to be eventually released in the stratosphere. The CFC lifetimes range from several to many decades.

2 HISTORY 1930-1986

Chapman[7] concluded that a catalytic agent would be necessary if an artificially induced attenuation of the ozone layer, for the benefit of uv astronomers, was to be attempted; he further observed that those involved would need to wear protective clothing. The next paper to deal with modifications to the ozone layer was Kofsky[8], who concluded that the ozone abundance would recover within 1000 seconds from total local photodissociation by the fireball radiation from an atmospheric nuclear weapon burst.

Harrison[9] considered the perturbing effect of water from the engines of a fleet of supersonic transport aircraft on the very dry stratosphere, via the OH chemistry then extant. This was soon shown to be a small perturbation on the global stratospheric water budget, even from a large SST fleet, but it paved the way for the milestone paper by Johnston[10], who argued that the nitrogen oxides produced by the aircraft engines could have very large effects. Several reviews and reports were produced[11-13], with the values estimated by numerical models of stratospheric photochemistry from a standard fleet fluctuating, even in sign, as the rate coefficients of important reactions in the O-H-N-Cl system were measured and remeasured. This phase largely ended in the 1980 s, developments in perturbation calculations since then having been mainly driven by the need to remedy deficiencies in the meteorology and radiative transfer of the numerical models of the stratosphere. The latest calculations[14] of the effects of high flying cruise aircraft have been triggered by renewed interest among manufacturers and NASA in SST's for transoceanic routes. The extent of NO_x emissions above 50mb (20km) is likely to be a crucial consideration. Some of the calculated results are shown in figure 1. Because of the intimate coupling between NO_x and Cl_x chain reactions, the effects of any aircraft fleet are calculated to be critically dependent upon the chlorine loading of the stratosphere, itself a strong function of past and future human activity. The aircraft-NO_x issue prompted estimates of the NO_x production by the atmospheric testing of nuclear weapons[15-18] in the late 1950 s and early 1960's. Large quantities were probably injected to the stratosphere, but it remains controversial as to whether or not an effect has been detected in the total ozone record; any reduction is certainly close to the natural noise level[19-21].

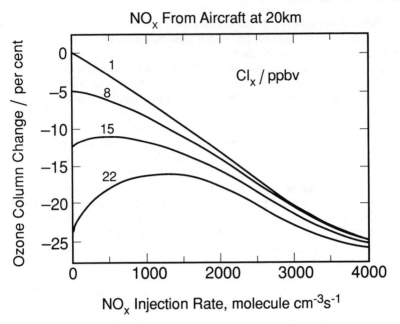

Figure 1. Percentage change of the vertical ozone column caused by supersonic aircraft, calculated by a one-dimensional model. The reference state had 1.1 ppbv upper boundary condition of Cl_x, with a lower boundary condition $N_2O = 300$ ppbv. For each nominal Cl_x condition (1, 8, 15, and 22 ppbv), nitric oxide injection rates of 0, 500, 1000, 2000 and 4000 molecules cm^{-3} s^{-1} at 20 km were evaluated. After Johnston et al[14].

In 1974 the focus switched to chlorine from chlorofluorocarbons, with the publication of the paper by Molina and Rowland[6], and has remained there ever since. Their original prediction, of high percentage ozone losses in the upper stratosphere of low and middle latitudes, has a very long time scale and still is not yet unequivocally evident in the data. However, chlorine-induced ozone loss became manifest in a totally unforeseen manner: in the lower polar stratosphere of late Antarctic winter and spring, where it was detected by Farman et al[22]. The decline proceeds at 1-3% per day within the vortex during late August, September and early October, resulting in 50% losses to the total column and up to 98% locally near 50mb. The total ozone column over Halley Bay (76°S, 27°W), Syowa (69°S, 40°E) and the South Pole during October is shown over the last three decades in figure 2; the phenomenon is evident from 1977 and is visible from year to year.

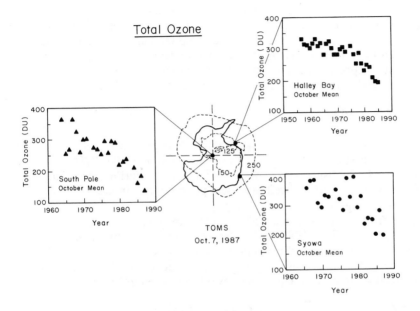

Figure 2. Total ozone losses, October means, from South Pole, Halley Bay and Syowa. The map shows the 125, 150 and 250 D.U. contours from the TOMS instrument on Nimbus-7, for 7th October 1987, which is the date of the lowest total ozone column ever recorded at the time of writing (February 1990). Source: Solomon et al[39].

Several hypotheses were put forward to explain the "ozone hole"; no one extant prior to the 1987 measurement campaigns had the complete picture, but crucial elements were foreseen. The role of polar stratospheric clouds[23] in converting chlorine from HCl and $ClONO_2$ to Cl and ClO via Cl_2, was suggested by Solomon et al[24], and their ability to irreversibly remove reactive nitrogen from the vortex by gravitational settling of mixed phase nitric acid-water crystals was suggested by Toon et al[25]. McElroy and his colleagues[26, 27] supported these suggestions and pointed out that bromine-based chemistry was also possible. Molina and Molina[28] suggested that the 3-body recombination of ClO to form Cl_2O_2 was the rate determining step, thereby making the ozone-destroying chain reaction dependent upon the square of the reactive chlorine abundance.

In the late 1970s it became apparent that stratospheric ozone was not immune from changes in the rates of emission from the surface of carbon dioxide[29]. This is also true of nitrous oxide[30] and methane[31], both of which are observed to be increasing. A comprehensive review of the entire subject of stratospheric ozone[32] was published in 1986, followed by an update[33] in 1990; they are highly recommended as source documents.

3 1986 - 1990

The last four years have been momentous. The ground-based campaigns to McMurdo (78°S, 167°E) in 1986/7 and to Thule (77°N, 69°W) in 1988 established that perturbed chemistry, featuring *inter alia* low NO_2 and high OClO, existed over both polar regions in the stratosphere during late winter and early spring. The airborne missions using aircraft from Punta Arenas (53°S, 71°W) in 1987 and from Stavanger (59°N, 6°E) in 1989 established that there was substantial denitrification induced by sedimentation of PSC particles, and that this condition permitted ozone loss at 1-3% per day, unequivocally attributable to chlorine chemistry. The sequence of events appeared to be:

$$ClONO_2(g) + HCl(s) \longrightarrow Cl_2(g) + HNO_3(s)$$
$$Cl_2 + h\nu \longrightarrow Cl + Cl$$
$$2[Cl + O_3 \longrightarrow ClO + O_2]$$
$$ClO + ClO + M \longrightarrow Cl_2O_2 + M$$
$$Cl_2O_2 + h\nu \longrightarrow ClO_2 + Cl$$
$$ClO_2 + M \longrightarrow Cl + O_2 + M$$

$$\text{nett:} \quad 2O_3 \longrightarrow 3O_2$$

with gravitational loss of $HNO_3(s)$ preventing the reformation of $ClONO_2$ via

$$HNO_3 + h\nu \longrightarrow OH + NO_2$$
$$NO_2 + ClO + M \longrightarrow ClONO_2 + M$$

There are many references to this work, which for the most part are in special issues of journals[34-39], and which all can be found in a recent review document[33].

One feature of the Antarctic ozone hole is the extent to which a single synoptic scale event, lasting for a few days as a tropospheric anticyclone extends polewards, can process large fractions of the vortical air through PSCs. This implies a link to climate, and large year-to-year variability in the hole. For the Arctic, one single event can have a major effect, as happened in 1989. The ozone losses in polar regions during winter raise the question as to whether or not they are transmitted to subpolar and mid latitudes, particularly in view of the northern hemisphere losses in the total column reported by the ozone trends panel[40], which are largest at higher latitudes during the winter months (figure 3).

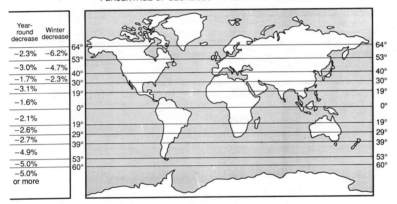

ATMOSPHERIC OZONE LEVELS: A GLOBAL DECLINE
PERCENTAGE OF OZONE LOST AROUND THE WORLD

Year-round decrease	Winter decrease
−2.3%	−6.2%
−3.0%	−4.7%
−1.7%	−2.3%
−3.1%	
−1.6%	
−2.1%	
−2.6%	
−2.7%	
−4.9%	
−5.0%	
−5.0% or more	

Figure 3. Percentage loss of column ozone. Between 30°N and 64°N the data are a combination from ground-based and satellite observations for the period 1969-86. Data from 60°S to the South Pole are from ground-based and satellite measurements since 1979. All other latitudes are obtained by satellite data alone from November 1978. Note that losses of 50% can occur over Antarctica in September and October. Source: Ozone Trends Panel, NASA RP 1208, 1988.

The Total Ozone Mapping Spectrometer (TOMS) produces global coverage, outside polar winter, and the maps of the trend in column ozone between 1978 and 1988 showed detectable decreases over the ten years[33], after the data were corrected for drift (see figures 4 and 5).

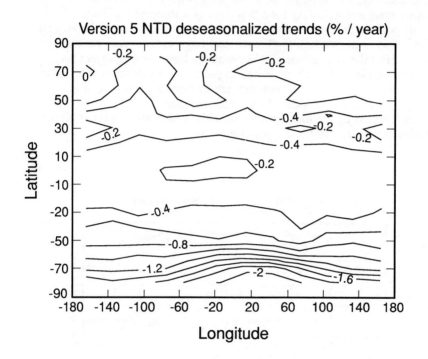

Figure 4. Linear fitted trend in daily total ozone measured by TOMS Nov 1978-Sep 1988. The contours are annually averaged trends in % per year fitted through deseasonalized data. The data are free of drift induced by diffuser plate degradation and use the V5 retrieval normalized to the Dobson network. Courtesy R.S. Stolarski and A.J. Krueger.

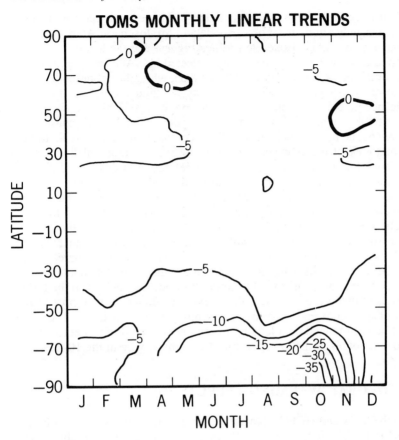

Figure 5. Linear fitted trend in monthly mean total ozone as measured by TOMS Nov 1978-Sept 1988, in %. Data are V5 retrieval normalized to the Dobson network. Note the occurrence of ozone loss in winter at subpolar latitudes outside the lower stratospheric winter/spring vortex, which is much longer lived in the Antarctic than in the Arctic. TOMS does not obtain data during periods of darkness in the polar night. Courtesy R.S. Stolarski and A.J. Krueger.

The information as to where in altitude these changes are occurring is less reliable, but indications from ozonesonde and Stratospheric Aerosol and Gas Experiment (SAGE) satellite profiles suggest that the lower stratosphere is involved[33]. The change detected by SAGE was -3±2% over 6 years at 25km, between 20° and 50° latitude, while the ozonesondes indicate -0.5% per year in the 17-24 km layer, a figure which is of limited significance globally because of station coverage. SAGE also shows -3% change at 40km, which is less significant for the total column because of the lower ozone number densities at 40 as compared to 25km. It should be noted that 80% of the ozone column is below 30km, a fact which would make ozone loss over the poles particularly effective in this regard if the so-called "containment vessel" of the polar vortex leaks.

The aircraft missions have produced data which have been interpreted, controversially, as showing that there is indeed mass flow through the Antarctic vortex, and moreover that there is ozone loss outside the vortex, even in August before the ozone hole proper is underway[41]. Others have argued, using the same data, that the rate of ozone loss over Antarctica in September demand "missing chemistry", even when complete containment is assumed. It is obviously vital to settle this question, and given the failure of models to predict the ozone hole, or even to simulate it well *post facto*, further observations and experimental campaigns are necessary. The same aircraft have produced data leading to arguments that during February 1989 ozone losses in the lower Arctic stratosphere were in the 10-20% range[38].

4 GLOBAL WARMING EFFECTS

Both CFC s and HCFC s can act as "greenhouse" gases, by absorbing infrared radiation, and may currently be contributing, when combined with N_2O, CH_4 and tropospheric O_3, as much as CO_2 to the calculated global mean surface warming[42]. The HCFC s are calculated to make much smaller contributions to the greenhouse effect than the CFC s which they are putatively to replace[43].

5 THE FUTURE

It is clear that models have limited prognostic power. This shortcoming is however balanced by the physical insight brought about by the recent field campaigns, particularly as regards polar ozone. The planned launch of the Upper Atmosphere Research Satellite (UARS) in late 1991 should enhance the global coverage of chemical species.

Completely halogenated hydrocarbons have, except for brominated molecules, long atmospheric lifetimes because their dominant sink is ultraviolet photo-dissociation at altitudes above much of the ozone layer in the lower stratosphere. Their slated replacements, HCFCs, have much shorter lifetimes because the hydrogen atom may be abstracted by reaction with tropospheric OH. A feel for the reduced effect calculated as being caused by the replacements in current models[43] is shown in figure 6. It should be emphasized that there are a number of important outstanding issues.

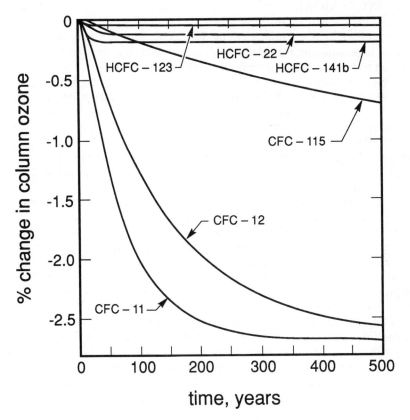

Figure 6. The time response of column ozone, following a step change of 5×10^8 kg a[-1] in the release of the specified gas. Redrawn from one-dimensional model calculations by Fisher et al[43]. Note that the calculations exclude the occurrence of polar ozone loss resulting from heterogeneous chemistry on polar stratospheric clouds, and cannot include the effects of large diabatic descent rates at high latitudes.

The very low values of CFC s, N_2O and CH_4 measured at 17-19 km in the polar vortices imply that numerical models are grossly underestimating the extent of diabatic descent there. In turn this implies a more rapid equator to pole circulation in the lower stratosphere. Two important factors could be affected by this: the recovery time of stratospheric ozone after cessation of CFC usage could be shortened, but the ozone-destroying ability of the HCFC replacements could be increased.

6 CONCLUSIONS

Over the past 20 years, numerical model calculations of future ozone losses from supersonic cruise aircraft, chlorofluorocarbons and the like have proved to be perishable commodities. Global satellite observations clearly show significant reductions in column ozone over the last decade, and the ground based network extends this back to 1969. Ascription of these trends is difficult in the absence of reliable prognostic models, but analysis of aircraft data has shown beyond doubt that the Antarctic ozone hole is caused by chlorine chemistry perturbed by reactions on polar stratospheric clouds. Less intense losses arising from the same processes are occurring in the Arctic. Given this, it is vital to establish whether the polar vortices are better approximated as containment vessels or flow reactors. The latter interpretation is consistent with the ozone losses shown in figure 5, which occur throughout winter equatorward of the jet stream maximum, that is to say outside the vortex.

It seems clear that winter and spring polar ozone loss, with its squared dependence on chlorine monoxide, will accelerate, even under the current version of the Montreal protocol. Replacement of CFC s by HCFC s holds out the promise of a return to pre-ozone hole amounts of stratospheric chlorine, on a time scale of decades. Should model calculations of the effects of HCFC s on stratospheric ozone prove to be underestimates, their shorter lifetimes would still permit more rapid corrective action than is possible with CFC s. The use of HCFC s should also lessen the "greenhouse effect" arising from CFC s. Continued measurement campaigns are necessary to improve understanding of the effects of CFC replacements.

Acknowledgements
    This paper has drawn on the multi-authored review document appearing in 1990 at the behest of WMO and other agencies. The author owes a large debt to the many scientists who participated in the Airborne Antarctic Ozone Experiment (AAOE) and the Airborne Arctic Stratospheric Expedition (AASE).

7 REFERENCES

1. S. Chapman, Mem. Roy. Met. Soc., 1930, 3, 103.
2. A.W. Brewer and A.W. Wilson, Q.J. Roy. Met. Soc., 1968, 94, 249.
3. J. Hampson, "Photochemical Behaviour of the Ozone Layer", Canadian Armament Research and Development Establishment (CARDE) Technical Note 1627, 1964, pages 11-18 and 266-267, and CARDE TN 1738, 1966.
4. J. Hampson, "Chemiluminescent Emissions Observed in the Stratosphere and Mesosphere", in Les Problemes Meteorologiques de la Stratosphere et de la Mesosphere, edited by CNES, Presses Universitaires de France, Paris,1965.
5. P.J. Crutzen, Q.J.Roy. Met. Soc., 1970, 96, 320.
6. M.J. Molina and F.S. Rowland, Nature, 1974, 249, 810.
7. S. Chapman, Q.J. Roy. Met. Soc., 1934, 60, 127.
8. I.L. Kofsky, J. Geophys. Res., 1962, 67, 739.
9. H. Harrison, Science, 1970, 170, 734.
10. H.S. Johnston, Science, 1971, 173, 517.
11. Climate Impact Assessment Program (CIAP), Monographs 1-5, 1975, U.S. Dept. of Transportation, DOT-TST-75-53, Washington, DC.
12. The Report of the Committee on Meteorological Effects of Stratospheric Aircraft (COMESA), Volumes 1 & 2, 1975, Meteorological Office, Bracknell.
13. National Research Council, "Environmental Impact of Stratospheric Flight", 1975, National Academy of Sciences, Washington, DC.
14. H.S. Johnston, D.E. Kinnison and D.J.Wuebbles, J. Geophys. Res., 1989, 94, 16351.
15. J.S. Chang and W.H. Duewer, "The Possible Effect of NO_x Injection in the Stratosphere Due to Past Atmospheric Nuclear Weapons Tests", AIAA Paper 73-358, AIAA/AMS International Conference on the Environmental Impact of Aerospace Operations in the High Atmosphere, Denver, Colorado, June 1973.
16. P. Goldsmith, A.F. Tuck, J.S. Foot, E.L. Simmons and R.L. Newson, Nature,1973, 244, 545.
17. H.S. Johnston, J. Whitten and J. Birks, J. Geophys. Res., 1973, 78, 6107.
18. J.S. Chang, D.J. Wuebbles and W.H. Duewer, J. Geophys. Res., 1979, 84, 1755.
19. J.K. Angell and J. Korshover, Mon. Wea. Rev., 1976, 104, 63.
20. G.C. Reinsel, Geophys. Res. Lett., 1981, 8, 1227.

21. P. Bloomfield, G. Oehlert, M.L. Thompson and S. Zeger, J. Geophys. Res., 1983, 88, 8512.
22. J.C. Farman, B.G. Gardiner and J.D. Shanklin, Nature, 1985, 315, 207.
23. M.P. McCormick, H.M. Steele, P. Hamill, W.P. Chu and T.G. Swissler, J. Atmos. Sci., 1982, 39, 1387.
24. S. Solomon, R.R. Garcia, F.S. Rowland and D.J. Wuebbles, Nature, 1986, 321, 755.
25. O.B. Toon, P. Hamill, R.P. Turco and J. Pinto, Geophys. Res.Lett., 1986, 13, 1284.
26. M.B. McElroy, R.J. Salawitch, S.C. Wofsy and J.A. Logan, Nature, 1986, 321, 759.
27. M.B. McElroy, R.J. Salawitch and S.C. Wofsy, Geophys. Res. Lett., 1986, 13, 1296.
28. L. T. Molina and M.J. Molina, J. Phys. Chem., 1987, 91, 433.
29. K.S. Groves, S.R. Mattingly and A.F. Tuck, Nature, 1978, 273, 771.
30. R.F. Weiss, J. Geophys. Res., 1981, 86, 7185.
31. R.A. Rasmussen and M.A.K. Khalil, J. Geophys. Res., 1981, 86, 9826.
32. "Atmospheric Ozone 1985", World Meteorological Organization, Global Ozone Research and Monitoring Report. No. 16, Vols I-III. Available from WMO, Case Postale No. 5, CH1211 Geneva 20, Switzerland.
33. "Scientific Assessment of Stratospheric Ozone: 1989", 1990, World Meteorological Organization, Global Ozone Research and Monitoring Project Report No. 20, Vols I & II. Available from WMO, Case Postale No. 5, CH1211, Geneva 20, Switzerland.
34. Geophys. Res. Lett., 1986, 13(12), 1191-1362.
35. Geophys. Res. Lett., 1988, 15(8), 845-930.
36. J. Geophys. Res., 1989, 94(D9), 11179-11738.
37. J. Geophys. Res., 1989, 94(D14), 16437-16860.
38. Geophys. Res. Lett., 1990, 17(4), 000-000.
39. "Scientific Assessment of Stratospheric Ozone: 1989", Vol I, Chapter 1, see reference 33.
40. International Ozone Trends Panel Report, "Present State of Knowledge of the Upper Atmosphere, An Assessment Report", 1990. Available NASA HQ, Code EE-U, Washington, DC 20546.
41. M.H. Proffitt, D.W. Fahey, K.K. Kelly and A.F. Tuck, Nature, 1989, 342, 233.
42. V. Ramanathan, L.B. Callis, R.D. Cess, J. E. Hansen, I.S. A. Isaksen, W.R. Kuhn, A. Lucas, F.M. Luther, J. D. Mahlman, R.C. Reck and M.E. Schlesinger, Rev. Geophys, 1987, 25, 1441.
43. "Scientific Assessment of Stratospheric Ozone: 1989", 1990, Vol II, Chapter IX, 381-404, see reference 33.

The Effects of Pollutants on Materials

R. N. Butlin
BUILDING RESEARCH ESTABLISHMENT, GARSTON, WATFORD WD2 7JR, UK

1. INTRODUCTION

One aspect of the effects of pollutants and acid
deposition which has been given much attention in recent
years is that on buildings and building materials.
However, until the 1980s little scientific research had
been undertaken in this area. Attention was drawn to this
by the House of Commons Select Committee on the
Environment in 1984[1]. A response to the Committee's
recommendations was a series of research programmes
carried out for the government by research organisations
and universities, and subsequently the publication of a
report entitled 'The Effect of Acid Deposition on
Buildings and Building Materials in the United Kingdom'[2]
in 1989 by the Building Effects Review Group (BERG).
Other international initiatives were also taken in the
1980s by the United Nations Economic Commission for
Europe (UNECE) following the publication of a review
covering wider aspects of the effects of sulphur
pollution[3]. Reference to current research programmes will
be made later in the paper.

As well as the effects of pollutants and acid
deposition on externally exposed materials there will be
other, sometimes secondary effects on artefacts in indoor
environments, for example objects displayed in museums or
art galleries. There is an increasingly extensive
literature covering this rather specialised area, and
reference to a recent review article by Brimblecombe[4] and
Thomson's book[5] on the Museum Atmosphere are good
starting points for references. This specialised topic
will not be covered in this paper nor will any possible

effects of acid deposition on water carrying
infrastructure (pipes, etc.) be discussed. Instead this
paper concentrates on the effects on materials exposed in
external building surfaces.

Since this paper is one of a series covering
different aspects of energy and the environment, and
since others will make detailed reference to trends in
air pollution concentrations and source attribution,
reference to these topics in this paper will be limited.
To put material effects into some kind of context
regarding emissions, it should be noted that average
smoke and sulphur dioxide concentrations in urban areas
of the United Kingdom have declined by factors of ten and
five respectively since the early 1950s. However, United
Kingdom emissions of oxides of nitrogen have roughly
doubled over the past 50 years. Since the 1950s, urban
concentrations of sulphur dioxide have decreased faster
than total emissions indicating that an increased
proportion now originates from emitters located outside
urban areas. In many areas vehicles, particularly those
burning diesel oil, are now the major source of smoke
concentrations.

Much of this paper is a condensed version of more
detailed discussion to be found within the BERG Report[2]
to which further reference should be made.

2. EVIDENCE FOR DAMAGE TO MATERIALS BY ACID
   DEPOSITION

2.1 Which pollutants are involved?

The primary pollutants which are usually considered
as leading to the degradation of building materials are
sulphur dioxide, combustion particles and chlorides.
Oxides of nitrogen may be involved either directly or as
part of a synergistic reaction. These primary pollutants
can also lead to the formation of equally damaging
secondary pollutants such as sulphates and nitrates.
Organic materials may also be affected by ozone. The
effect of these pollutants must be seen against the
'background' damage that will occur as a result of
natural weathering phenomena such as moisture, frost,
marine aerosols, etc. Materials that were exposed to
higher levels of pollution in the past may continue to
show some effects as a result of the persistent reaction
products absorbed into the surface. Estimations of the
historic levels of pollution in York have been discussed

by Brimblecome[6]. In addition, the the type of deposition
(wet, dry, or occult) and the meteorological variable
such as temperature, humidity, wind speed and direction,
etc. will also affect rates of reaction of pollutants
with materials and consequently these must be taken into
account when assessing damage or quantifying effect.

2.2 Evidence for damage to stone

Various studies have been made of the weathering of
limestone in standing buildings (Table 1). A
comprehensive study of weathering of the stone,
measurement of wet and dry deposition, and rainfall
'run-off' from St.Paul's Cathedral, London[7] showed that
the mean hydrogen ion concentration of incident rain
water was equivalent to pH 3.8 to pH 4.9. As the rain
water passed over the stone the hydrogen ion
concentration decreased and the total hardness increased.
The material that was eroded as sediment was derived from
the area where the rain first landed rather than the
areas it drained across. Estimates of the surface erosion
(in solution and as sediment) on the balustrade on the
roof of St.Paul's Cathedral range from $220\mu m$ yr^{-1} to
$139\mu m$ yr^{-1}. The mean erosion for the period 1727 to 1982,
based on the differential erosion of the stone and the
lead plugs in the stone (sealing the lifting holes) is
$78\mu m$ yr^{-1}.

Other evidence of stone decay comes from site
exposure trials. Results from samples exposed at Garston
(Hertfordshire) and Westminster[8] between 1955 and 1965
showed weight losses equivalent to a minimum surface
recession rate of $23\mu m$ yr^{-1} at Westminster and $10\mu m$ yr^{-1}
at Garston. Another study was carried out by NATO/CCMS[9]
in a number of countries including the U.K. The samples
of stone (from Germany) exposed at St.Paul's Cathedral
showed minimum surface recession rates of $33\mu m$ yr^{-1}. Some
correlations were found between the atmospheric pollutant
concentrations measured using a passive monitor and the
pollutant reaction products in the stone.

Data for rates of 'natural' or background weathering
over thousands of years have been obtained from outcrops
of carboniferous limestone. The rates are equivalent to a
surface recession rate of between $4\mu m$ and $75\mu m$ yr^{-1}.
Modern weathering rates are higher than most of these
background rates which confirms that elevated pollution
concentrations increase the rates of degradation of
stone. It is known that carbonic acid (carbon dioxide

Location	Stone Type	Decay Rate	Period	Reference
London (St.Pauls)	Portland Limestone	220μm yr^{-1} (Run-off anal.)	1980-81	7
		139μm yr^{-1} (Direct meas. of erosion)	1980-82	7
		78μm yr-1 (Differential erosion)	1727-1982	7
London (St.Pauls)	Baumberg Sandstone	4.0% yr^{-1}	1980-82	9
London (St.Pauls)	Muschelkalk Limestone	3.5% yr^{-1}	1980-82	9
Garston, Herts	Portland Limestone	0.13% yr^{-1}	1955-65	8
London (Whitehall)	Portland Limestone	0.33% yr^{-1}	1955-65	8
South-East England	Portland Limestone	0.46% yr^{-1} (urban)	1981-83	10
		0.34% yr^{-1} (rural)	1981-83	10

Table 1. Measured decay rates for stone exposed in the United Kingdom.

dissolved in water) makes a contribution to the degradation of stone, but the studies so far are limited and have not shown how significant this process is relative to sulphur dioxide.

2.3 Evidence for Damage to Metals

The literature on the atmospheric corrosion of metals is voluminous and will not be referred to in detail in this paper. Many of the early site trials on the corrosion of metals did not include quantitative information on pollution associated with each site. Some measurements of rainfall constituents and sulphur dioxide were made in the 1930s but generally relationships between corrosion and pollutant levels were not derived. In the last 30 years, few extensive surveys of corrosion have taken advantage of improved pollution measuring capabilities.

Corrosion of iron and steel is primarily caused by oxygen and moisture, accelerated by contaminants such as sulphur dioxide, particulates and chlorides. Addition of copper, nickel or chromium reduce the corrosion susceptibility of the steel (Table 2). Few relationships between damage, pollutants, and meteorological conditions have been determined from U.K. studies. Studies outside the U.K. have derived relationships which differ in form, variables, and in the definition of 'time-of-wetness'. Observed rates of corrosion[11-12] include 48μm yr^{-1} for ingot iron (Surrey) to 173μm yr^{-1} (Derby). The presence of 0.3% copper in the steel slowed the rate[11-12] to between 36 and 109μm yr^{-1} in the same trials. Rates of corrosion of copper bearing steel at Stratford (London) were 62.5μm yr^{-1} in 1978-80 and 37.6μm yr^{-1} in 1983-84, in line with reduced sulphur dioxide levels. In the period 1930-60 rates in the same area were typical 100μm yr^{-1}.

In the atmosphere, zinc corrosion rates are slow[11-12] and constant at around 1 - 10μm yr^{-1}. Copper and its alloys are highly resistant to atmospheric

Metal	"Rural" corr.rate (μm yr^{-1})	"Urban" corr.rate (μm yr^{-1})	Ref.
Ingot Steel	48	173	11,12
Copper Steel	36	109	11,12
Copper Steel	–	62	BSC 1979-80
Copper Steel	–	38	BSC 1983-84
Zinc	1	10	11,12
Copper	0.2-0.6	0.9-2.2	13
Aluminium	–	2-5 (surface) 250-500 (pitting)	14
Tin	0.05	0.125-0.175	15
Magnesium	–	76	16
Nickel	0.25	1.75	17

Table 2. Measured corrosion rates for various metals in "rural" and "urban" pollution environments.

corrosion because of the surface layer of cuprous oxide
plus basic salts. Penetration rates range[13] from 0.2 -
0.6μm yr^{-1} (rural) to 0.9 - 2.2μm yr^{-1} (industrial). Lead
and its alloys are also highly resistant to corrosion but
performance in chloride contaminated atmosphere may be
slightly impaired. Aluminium alloys have good resistance
to atmospheric corrosion. Rates observed[14] in industrial
environments are 2 - 5μm yr^{-1}. However, some pitting to a
greater depth takes place. Tin corrodes very slowly in
the atmosphere and magnesium alloys are essentially
immune at relative humidities below 60%[15]. Nickel is also
highly resistant to atmospheric decay[17].

Evidence for the decay of metals and non-metals
shows all the features expected if atmospheric pollution
is accelerating weathering and damage. Evidence is not
unequivocal that present rates of decay on historic
buildings are less than those of the past. Evidence for
metals shows a reduction in the rate of weathering for
steel but little for other metals. This may, however, be
challenged by current studies (see below). There is no
evidence as to whether 'safe' levels exist or which
atmospheric components (moisture, frost, salts, sulphur
dioxide, oxides of nitrogen, chloride, carbon dioxide,
etc.) are the major causes of damage.

3. ESTIMATION OF BUILDINGS AND MATERIALS AT RISK

Some materials are very susceptible to pollutant damage,
others are less susceptible but are more often used. For
example, concrete is only slightly damaged by acid
deposition but the financial importance of damage could
be increased by the large quantities involved. Metals,
calcareous stones (limestones, magnesian limestones and
calcareous sandstones) are likely to be the worst
affected by acid deposition. Other materials are damaged
by 'natural' processes but not necessarily worsened by
acid deposition. One exception may be medieval glass.
It is known to be damaged by moisture and the decay
products include sulphates[18] that behave as a hygroscopic
layer that leads to further decay. Organic materials
(especially those with inorganic fillers), brickwork,
cement, concrete, and non-calcareous slate may be
affected. Organic growths will also affect the weathering
of stones, though their growth can also be hindered by
high levels of pollution.

In a building the location of a particular material
in the structure will affect the damage it may suffer[19]

and external features of buildings can be 'zoned'
according to the weathering of the stone in those areas.
Problems and damage to materials can also be exacerbated
by the relative positions of different materials (e.g.
stone and metal lead to staining or cracking of the
stone).

The establishment of an inventory of buildings and
their constituent materials is difficult. To estimate the
damage to materials from acid deposition the total
quantities of exposed surface area of susceptible
materials must be estimated or calculated. This can be
done either by an inventory/census approach where each
building is examined individually, or by a probability
distribution approach where all areas of exposed
materials are aggregated to provide a probability
function of the exposed area per unit area of land. The
former method is appropriate to cultural buildings which
are essentially unique (e.g. cathedrals or churches).
Some lists of such buildings are available from heritage
bodies.

Recent studies in the U.K. using the probability
approach have concentrated on developing a
'stock-at-risk' methodology which estimates the total
quantity (surface area) of exposed materials and
redefines 'stock-at-risk' via building components. It
involves deriving 'identikits' for different types of
buildings with each type specified by number of stories,
floor space, basal surface index, number of buildings per
km^2, and age. The estimation of the external envelope is
made and field surveys used to define typical materials
used in a construction type. Land use maps are then used
to determine the proportion of each km grid square
occupied by particular building type. The methodology
has been applied to cities in the U.K. and in Germany.
The estimates derived by such methods are being used in
calculations of economic benefits of reducing pollutant
emissions.

4. MECHANISMS OF DEPOSITION AND DECAY

4.1 Deposition

Deposition of pollutants (particulates, aerosols or
gases) onto material surfaces is generally defined as
either wet or dry. There is more information available on
the wet and dry deposition of sulphur species than any
other in the U.K. The dry deposition of pollutants, gases

or particulates onto surfaces from the atmosphere is a
complex process involving both atmospheric and surface
processes. A concept of surface resistance comprising
different factors has been used which enables a
'deposition velocity' to be determined. This represents a
measure of the amount of a pollutant species deposited on
a surface relative to the atmospheric concentration.
Aerodynamic resistance (determined by a combination of
wind velocity and aerodynamic roughness of surfaces)
appears to be the principal factor governing dry
deposition of sulphur dioxide in a city. The dry
deposition of oxides of nitrogen is less well understood
than that for sulphur dioxide, especially in urban areas.
One general problem is that those deposition velocities
which have been determined are averages for all surfaces
including vegetation in urban areas. Consequently, data
specific to buildings are not available without further
research.

    Pollutants can become incorporated into
precipitation (wet deposition) either by in-cloud
scavenging ('rainout') or below cloud scavenging
('wash-out') with the former being the more important. In
areas remote from major pollution sources and with high
rainfall, wet deposition will predominate. Dry deposition
will be dominant in Eastern England. An estimate of the
relative importance of wet and dry deposition processes
on the erosion of stone tablets indicated that up to 40%
could be caused by dry deposition[10]. Both wet and dry
deposition will be affected by wind velocity and rain
intensity which are both particularly variable around
buildings in urban areas. Surface features on a building
can also play a significant part in the course of
deposition and in the sheltering of certain parts from
wet deposition.

4.2 Damage Mechanisms

    4.2.1 Stone. The 'natural' weathering of porous
stone can include frost damage and impregnation with
soluble salts many of which hydrate and expand on
crystallisation within the pore structure. In polluted
atmospheres, calcareous stones can be affected by sulphur
dioxide via the formation of calcium sulphate from the
carbonate. As the sulphate is more soluble than the
carbonate it will either wash off in the rain – resulting
in a weight loss or recrystallise and hydrate within the
pores causing subsequent disruption of the surface. This
latter process may occur over a long period of time. The
pH range of wet deposition will affect the dissolution

process as will transport and diffusion of reactants to the surface. The roles of oxides of nitrogen and carbon dioxide in the overall decay processes of susceptible stones is still not understood. Rain-sheltered stone is often covered in crusts that are blackened by soot. These crusts are rich in hydrated calcium sulphate and form by dry deposition of sulphur dioxide into the pores of moist stone. There is little evidence of nitrate formation in the crusts.

Particulate matter, from the combustion of coal or oil, is often found on the surface of decayed stone. The particles contain carbon, silicon, sulphur, aluminium and copper as major constituents and may either have sufficient sulphur in them to give rise to gypsum on the stone surface[20], or may catalyse the oxidation of sulphur dioxide to sulphate[21]. Organic species in the form of algae, lichens and possibly bacteria may also have a definite role to play in the erosion of stone samples[22].

4.2.2 Metals. Atmospheric corrosion of metals is known to be an electrochemical process relying on the presence of water or water vapour, a corrosion stimulating agent and oxygen[23]. The thickness and transitory nature of the moisture film on the metal surface is characteristic of atmospheric corrosion. Layers of oxide produced in dry conditions break down when exposed to water and permit a much more rapid electrochemical reaction, the product of which does not usually form a coherent barrier to further attack. Relatively low rates of corrosion in wet conditions are affected by strong electrolyte solutions which dissolve the oxide film. Rain water contaminants such as sulphuric acid, hydrochloric acid, sea salt, atmospheric sulphur dioxide, and ammonium salts can contribute to the formation of corrosive solutions. Relative humidity plays a significant part in the corrosion of metals as does the time during which the metal is wetted. Maximum rates occur when the water film is at its thinnest immediately before drying. Particulate solids such as ammonium sulphate, chlorides on iron in the presence of sulphur dioxide can initiate localised corrosion and increase the aggressivity of the sulphur dioxide.

4.2.3 Other Materials. There is little evidence that pollutants enhance damage to brickwork, the main decay mechanism being the transfer of sulphate in the brick to the mortar resulting in the expansive and disruptive formation of the mineral ettringite. The major effect of the atmosphere on reinforced concrete is through the

absorption of carbon dioxide which destroys the passive
alkaline protection around the steel reinforcement and
gives rise to corrosion. Sulphur dioxide in the
atmosphere is not thought to have a major effect on
reinforced concrete. Medieval glass is less durable than
modern glass because the high levels of calcium and
potassium ions suppress the silicon content. The removal
of calcium or potassium from the surface by corrosion
results in a less durable structure than that of modern
glass. Sulphate is often found in the corrosion layer but
the initial corrosion occurs as a result of the formation
of hydroxides followed by carbonation and then
sulphation. The sulphates form a hygroscopic crust on the
surface of the glass which promotes attack by prolonging
surface wetness[18].

4.3 Conclusions

The interactions between building materials and
pollutants are very complex. Deposition of pollutants
onto surfaces depends on the atmospheric concentration,
the strength and direction of the wind and on the rain
intensity. Interactions of pollutants once on the surface
will vary with the position of the material, the
time-of-wetness, and the structure, reactivity and degree
of protection of the material involved. Calcareous
stones, iron, steel, zinc and medieval glass are at
greater risk than other materials. Water, chloride, and
sulphur dioxide (and associated salts) are important in
accelerating damage relative to 'natural levels'.

5. PREDICTING DAMAGE

Damage functions are the quantitative relationships
between pollutant concentrations, meteorological
variables and the rates of decay of particular materials.
Rates of decay or corrosion depend a range of factors
including 'time-of-wetness' (duration of rain, number of
days with rain, etc.), the supply of gaseous pollutants,
the supply of pollutants in solution in rain-water,
ammonium salts, organic acids and particulates. Other
factors are temperature and time of day during which
water in contact with the surface is frozen. Free
exposure and sheltered (rain excluded) conditions will
also affect the rates of decay. The development of
predictive equations must be based on a wide range of
measurements of the composition of the atmosphere, of
rain water and other factors from a series of exposures
to a wide spectrum of pollution climates. For

completeness not only should average rates of decay be considered against average values of pollution levels, but transient characteristics of decay processes should be investigated along with measurement of transient changes in the atmosphere. Equations published to date usually only involve an average sulphur dioxide concentration and a 'time-of-wetness' related term, and are derived from tests that did not include measurement of chlorides, oxides of nitrogen and other species[25]. Alternatively they can be derived from laboratory experiments that bear no resemblance to atmospheric conditions. In general, available damage functions are not capable of reliably predicting rates of corrosion for samples not included in the original data set from which the function was derived.

In order to redress this dearth of information, a National Materials Exposure Programme has been set up in the U.K. to determine rates of decay in current pollution climates. A series of metal and stone samples (Table 3) have been exposed at 29 sites where rainfall, temperature, humidity, pH, sulphur dioxide, nitrogen dioxide, black smoke, and wind velocity are all measured. This programme involves BRE, National Power, British Coal Corporation, Warren Spring Laboratory and the National Physical Laboratory has been running for three years and the first stage will be completed in 1991. The data from this programme is being used to create a database and as the basis of mathematical models to explain the observed decay patterns.

Perhaps the main question being addressed by the NMEP is the effect of different reductions in pollution levels (e.g. reducing sulphur dioxide by flue gas desulphurisation) on buildings and building materials, especially on the costs of repair, etc. Modelling of results using basic scientific theories should enable more widely applicable predictions to be made for areas of the U.K. not covered by the programme. A similar exposure programme is being run by the UNECE in 12 countries. The materials exposed are similar to those in the NMEP but also include bronze, wood and electrical materials.

All Sites: Mild Steel

              Aluminium

              Galvanised Mild Steel

              Alkyd Paint System
              (on mild steel)

              Copper

              Portland Limestone

              White Mansfield
              (dolomitic sandstone)

              Monk's Park Limestone
              (BRE sites only)

Table 3. Materials in the National Materials Exposure
         Programme.

6. ECONOMIC COST OF DAMAGE TO MATERIALS

Given adequate knowledge of the relationships between
atmospheric variables and rates of decay of materials,
and an idea of the 'stock-at-risk' then it is possible to
make an assessment of the cost of the damage and
consequently the benefits from reduced damage (e.g. costs
of maintenance cycles, and service life of components).
However, in the absence of adequate damage functions the
task is difficult. Attempts have been made to relate
costs of damage to pollutant concentrations by comparing
expenditure on repair and replacement in clean and
polluted areas. But more commonly the costs are based on
a damage function and are derived from the damage
calculated for each material studied, the unit cost of
repair or replacement of materials and the amount of
material exposed. There is much uncertainty at each stage
of the computation and this must be taken into account.
Calculations of costs of damage for modern buildings are
in practice easier than for historic buildings where
benefits and 'replacement' costs are harder to quantify
and other methods need to be used.

7. PROTECTION AND PRESERVATION OF BUILDING MATERIALS

7.1 Introduction

Most buildings materials will be damaged by weathering even in the absence of pollution. Many deleterious effects are, however, accelerated by anthropogenic pollutants. Treatments can be used either to protect new components or to prolong the life of components that are already damaged or weathered.

7.2 Metals

These are used for structural strength, as sheeting, for channel sections and pipes, and for decorative features. Problems of metal corrosion are well understood and economic protective measures are available providing provision is made at the design stage.

Plain carbon steels have poor resistance to corrosion in open exposure and therefore must be protected. Low alloy steels have better resistance with stainless steels requiring the least protection. The commonest form of protection for steelwork is painting which not only excludes oxygen and water from the surface but also excludes strong electrolytes. Direct attack by atmospheric agents on paint work can result in cracking, blistering and general thinning. Decay of the substrate can damage or detach the paint film. There is little indication that sulphur dioxide plays a major part in damage to paint although drying may be impeded and pigments may be affected if based on calcium carbonate. Polymeric materials, metal coatings and ceramics can also be used to protect metals. Preservation requires comprehensive removal of corrosion products before re-treatment.

Zinc is comparatively reactive and in combination with iron sets up an electrochemical cell that protects the iron from attack whilst corroding the zinc 'sacrifically'. In terms of protection and preservation it is often thought to be more cost-effective to allow galvanised steel to approach failure and replace it rather than apply extra protection, especially in the case of inexpensive items. The resistance of copper to atmospheric corrosion is better than that of zinc and much better than steel. Corroded surfaces are protected by a layer of sparingly soluble hydrated salts. Protection and preservation treatments are not usually required. Lead also has good resistance to external

atmospheric corrosion and can be painted although this is
not necessary for protection. Aluminium and its alloys
are highly resistant to atmospheric attack . The rate is
somewhat accelerated in industrial areas but still very
much slower than steel. Anodising aluminium thickens the
oxide film that naturally protects the surface. Even
greater protection can be achieved by the addition of
corrosion inhibitors to this layer.

7.3 Non-Metals

The effect of atmospheric pollution on wood is very small
compared to that of water and humidity. Wood can be
protected by fungicides, biocides, paints, oils, waxes,
and resins.

     There are no satisfactory ways of protecting stone
from atmospheric attack in the long term. Some
preservatives have been developed especially for the
consolidation of badly damaged stone, but there is doubt
in some cases about long term effects on the basic
materials. Some natural slates containing carbonates are
susceptible to atmospheric attack but there is no record
of conservation techniques being used.

     Anti-carbonation coatings are being developed for
concrete, but as stated earlier, the effects of
pollutants (apart from chlorides) are insignificant
compared with the effect of carbon dioxide.

     At present the main method of protecting or
preserving valuable glass is to install externally
ventilated glazing in front of the existing glass.
However, some current studies of coating for medieval
glass are taking place.

     8. CONCLUSIONS

Pollutants generated during energy production will have
some effect on materials in the external envelope of a
building. The more susceptible materials will be metals
and calcareous stones and the degree of attack will
depend on the ambient pollution climate. Knowledge of
rates of attack and mechanisms of decay varies widely
with the different materials. But in general there is a
dearth of information on damage functions and the
quantitative relationships between pollutant
concentrations, meteorological variables and rates of
decay. National and international programmes have been

set up to determine the current rates of decay for a number of materials. The exposure programmes are being backed up with a range of laboratory studies and research into particulates, gypsum crusts, historic levels of pollution, and studies of individual buildings. All this must be put in the context of the natural background decay of materials and the rates of weathering in the recent historic past.

The main thrust of a comprehensive assessment of the effect of pollutants on materials must the calculation of the economic benefit of reduced levels of pollution. This requires a means of assessing the 'stock-at-risk' and methods of assessing the benefits in economic terms. There have been useful developments in both these fields recently. Other relevant factors in the general debate include the role of 'episodes' of pollution, the source allocation of pollutants ('near' versus 'far', 'industrial' versus 'domestic'), the existence of threshold levels or critical loads for different materials, none of which are discussed in detail here. Nevertheless, progress has been made in the last four years in these areas and results will be presented at forthcoming symposia and in the open literature.

REFERENCES

1. House of Commons Environment Committee. 'Acid Rain. Fourth Report', Session 1983-84. HMSO, London, 1984.
2. Building Effects Review Group, 'The Effects of Acid Deposition on Buildings and Building Materials in the United Kingdom', HMSO, London, 1989.
3. United Nations Economic Commission for Europe (UNECE), 'Air-borne sulphur pollution - Effects and control report prepared within the framework of the convention on long range transboundary air pollution', United Nations, New York, 1984.
4. P. Brimblecombe, Atmos.Environ., 1990, 24B, 1.
5. G. Thomson, 'The Museum Environment', Butterworths, London, 1978.
6. P. Brimblecombe, 'The Big Smoke - A History of Air Pollution in London since Medieval Times', Methuen, London, 1987.
7. R.N. Butlin et al., Proc.Vth International Congress on Deterioration and Conservation of Stone (Lausanne), 1985, 537.
8. S.M. Jaynes, 'Studies of Building Stone Weathering in South-East England', Unpublished Ph.D Thesis, University of London, 1985.

9.   North Atlantic Treaty Organisation, 'Atmospheric measurements of air pollution – Pilot study on conservation and restoration of monuments', Report Aff/S 598–85, No. 158, NATO, Germany, 1985.
10. S.M. Jaynes and R.U. Cooke, Atmos.Environ., 1987, 21, 1601.
11. BISRA, Iron and Steel Special Report 21, 1938.
12. BISRA, Iron and Steel Special Report 66, 1959.
13. E. Matson and R. Holm, American Society for Testing Materials Special Technical Paper, 435, 187, 1968.
14. F.L. McGeary et al., ASTM STP, 435, 141, 1968.
15. S.C. Britton, International Tin Reasearch Institute Pub. 510, 1976.
16. L.L. Shier, 'Corrosion', Butterworth, London, 1976.
17. V.E. Carter, 'Corrosion Processes' (Ed. R.N. Parkins), 77, 1982.
18. R.G. Newton, 'The deterioration and conservation of painted glass: a critical bibliography', Oxford University Press, Oxford, 1982.
19. E. Leary, 'The Building Limestones of the British Isles', HMSO, London, 1983.
20. M. Del Monte et al., Atmos.Environ., 1981, 15, 654.
21. J.B. Johnson et al., J.Electrochem.Soc., 1983, 130, 1650.
22. A. Ciarallo et al., Proc.Vth International Congress on Deterioration and Conservation of Stone (Lausanne), 1985, 607.
23. U.R. Evans, 'The Corrosion and Oxidation of Metals', Arnold, London, 1960.
24. W. Muller-Golchert, Mappe, 1982, 210, 772.
25. F.W. Lipfert, Atmos.Environ., 1989, 23, 415.

ACKNOWLEDGEMENTS

Contributed by courtesy of the Director, Building Research Estsblishment, U.K. and reproduced by permission of the Controller, HMSO: Crown copyright 1990.

The Environmental Impact of the Nuclear Fuel Cycle

L. E. J. Roberts
ENVIRONMENTAL RISK ASSESSMENT UNIT, UNIVERSITY OF EAST ANGLIA,
NORWICH NR4 7JJ, UK

1 INTRODUCTION

The fission of uranium is a very concentrated source of
energy. A nuclear power station of the type we use today
uses about 200 t of fresh uranium fuel to generate 1 GW,
1000 MW, of electricity. An equivalent coal-fired
station would use about 3.7 million tons of coal a year.
The transport and storage of nuclear fuel is therefore a
much smaller operation and nuclear power plants do not
have to be sited close to the source of fuel. Waste
disposal is also a much smaller-scale problem than in the
case of coal. Further, power from uranium fission
generates no atmospheric pollutants like SO_2 or NO_x. A
small amount of CO_2 would be produced by the
transport and diesel fuel needs of different parts of the
nuclear fuel cycle, but this would be at most a few
percent of the CO_2 produced in any cycle burning coal, oil
or gas. Nuclear power stations, like any other power
stations, are large-scale construction projects, needing a
large supply of cooling water but taking rather less land
than fossil fuel stations because there is no need for
fuel stores. But the fuel cycle makes comparatively
little environmental impact.

However, the fission process also results in the
production of radioactive materials and the environmental
impact of the nuclear fuel cycle concern the effects of
any discharge of radioactive materials, either a planned
discharge or an uncontrolled release due to an accident.
The purpose of this paper is to give a broad overview of
the extent and possible effects of the release of

radioactive materials from nuclear fuel cycles.   It is a
subject which has attracted an enormous amount of
attention and research.

The Control of Radiation Levels

      Ionising radiation is the hazard which is associated
with nuclear power and particularly the effects on people.
A great deal of work has indicated that, if humans are
adequately protected against radiation, the effects on
other species will also be low.   The industry is strictly
regulated.   Radiation doses to workers and to the public
are subject to government regulations, advised in this
country by the National Radiological Protection Board
(NRPB) and internationally by the International Commission
on Radiological Protection (ICRP).   As far as risks of
accidents are concerned, the safety standards in the
industry are set by the Health and Safety Executive,
working through the Nuclear Installations Inspectorate
(NII).   Every nuclear installation must be licensed by
the NII and is subject to inspection by them.   The HSE
has recently published a paper for consultation on "The
tolerability of risk from nuclear power stations".[1]

      The radiological regulations set maximum limits, and
also require management to reduce doses below them to
levels As Low As Reasonably Achievable, the ALARA
principle.   The NRPB have recently recommended lowering
the maximum allowable annual doses for workers to 15 mSv*
and to the public to 0.5 mSv.[2]   These are the additional
doses due to exposure to ionising radiation from an
industrial source and are an addition to the background
radiation to which we are all subject, on average 2.2
$mSvy^{-1}$ in the UK.   Natural background varies considerably
across the country, from 1-100 mSv/y,[3] mainly due to the
variation of the dose from indoor radon;   e.g. the average
annual radiation dose in Cornwall is 7.8 mSv.[4]   Some
50,000 people in the general population have been living
with annual doses above 20 mSv, mostly due to indoor radon

Footnote
* The Sievert (Sv) is the unit for the "effective dose
equivalent", which is a measure of the total risk to
health from the energy absorbed in the body from ionising
radiation.   The more practical unit is the milliSievert,
10^{-3} Sv or mSv.   "Effective dose equivalent" is condensed
to "dose" in this paper.

levels.   Some typical values of radiation dose regulation
and exposure are collected in Table 1.

The application of the ALARA principle has been
successful in reducing the average exposure below maximum
levels.   The annual average dose to workers in the
nuclear industry is 2.0 mSv above background, and 914
workers   received   above 15 mSv in 1987, as did 500 miners
in non-coal mines.[3]   The most highly exposed group of the
public to nuclear discharges is a group eating fish
locally landed near Seascale;   their excess dose due to
discharges from Sellafield was 0.35 mSv in 1988, while the
excess dose due to the natural radionuclides in fish
would be 0.43 mSv.[4]

The effects of radiation dose have been studied by
many bodies, including the ICRP and the United Nations
Scientific Committee on Atomic Radiation (UNSCEAR).
Acute radiation syndrome only sets in at very high doses,
above 1 Sv delivered in a short time.   At lower doses,
there is an enhanced risk of cancer, and of damage to the
foetus in certain stages of pregnancy.   The latest advice
from the NRPB is that the   danger   of contracting a fatal
cancer is approximately 4.5×10^{-5} per mSv absorbed.[2]
The range that   can be   deduced from the latest UNSCEAR
report is $0.7-3.5 \times 10^{-5}$ per mSv of $\beta-\gamma$ at low doses.   On
the usual assumption that the risk is linearly
proportional to dose, the risk accruing to a population
may be estimated from the sum of all doses received by
individuals, or by multiplying the average dose by the
number of people;   this is called the "collective dose"
and measured in man Sv.   The collective annual dose to
the UK population is then 137,500 manSv.   Application of
the NRPB factor would imply that about 6000 of the 150,000
fatal cancer cases in the UK every year may be due to
background radiation.   The total collective dose to the
UK population due to discharges from the nuclear industry
is 30 manSv, less than 0.1% of the total, and would
correspond to an extra 1-2 cases a year.

Studies of the incidence of cancer in workers in the
AEA and in BNF plc have not revealed any general increase,
except possibly for prostatic cancer in the case of a
small number of workers exposed to more than 100 mSV and
to internal contamination.[6,7]   In general, the number of
cancer-deaths recorded was less than the average for the
population.

Table 1   Examples of Radiation Dose (milliSieverts)

1000 mSv   Threshold above which acute sickness possible.
 100 mSv   Maximum annual dose due to radon in houses.
  50 mSv   Maximum permitted annual dose for workers.
  15 mSv   Recommended maximum annual dose for workers.
  10 mSv   Recommended maximum dose due to indoor radon.
  2.2 mSv   Average annual dose due to background (UK).
  2.0 mSv   Average annual excess dose to nuclear workers.
  2.0 mSv   Annual average dose to aircraft crew.
  0.5 mSv   Recommended maximum annual dose to the public
            from single nuclear installations.
  1.2 mSv   Average dose from one X-ray of the pelvis.
  0.3 mSv   Annual average dose due to medical uses.

An allowance has also been made in radiological
protection limits for a possible result of genetic damage
due to radiation, though the risk is assumed to be lower
than the risk of contracting cancer.   In fact, the risk
in this case has been deduced solely from experiments on
animals, since no human cases were found even in
populations that had received considerable irradiation.
However, a recent report records an association between
cases of leukaemia in people less than 25 years old and
exposure of their fathers to occupational doses in
Sellafield.[5]   Of 46 cases in West Cumbria, 8 had fathers
who had been employed at Sellafield and had a recorded
radiation dose.   An exposure of 100 mSv or more (4 cases)
was associated with a six- to eight-fold increased risk of
leukaemia in the offspring.   The risks have large
uncertainties associated with them, and increased risks
were also recorded for the children of steelworkers,
farmers and workers in the   chemical industry.   More
studies are urgently required.[6]   However, the effect, if
it is real, seems to be associated with exposure levels
well above those suffered by the general population.

2   NUCLEAR FUEL CYCLES

A typical fuel cycle for a large, modern AGR or PWR, using
oxide fuel enriched to 3% U-235, is shown in Figure 1.
The quantities of materials are approximately those that
would be involved in generating 1 GW of electricity for
one year, 1 GW(e)y.   The first generation of nuclear
reactors in Britain, the gas-cooled "Magnox" reactors,
used natural uranium metal fuel clad in a magnesium alloy.
In such a case, the enrichment stage is omitted and the

turnover of fuel would be approximately 200 rather than 37 t.   The quantities of material in the fuel cycle are low, because of the concentration of energy in nuclear fuel already alluded to, and the only operation involving handling a large bulk of material is the mining of the original uranium.   A typical ore contains 0.05 to 1% of uranium, and the provision of 200 t of uranium requires mining 20,000 to 400,000 t of ore and the waste rock associated with it.

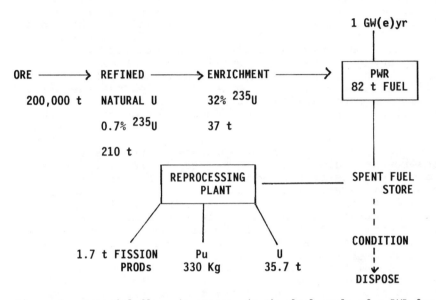

Figure 1   Material flow, in tonnes, in the fuel cycle of a PWR for the generation of 1 GW(e)y.

     The major source of radioactivity in the whole fuel cycle is the spent fuel itself, due to the radioactivity of the fission products.   Spent fuel must therefore be discharged into cooling ponds, though gas or air cooling can be substituted later.

     A critical decision is whether or not to reprocess the spent fuel, to recover the uranium, depleted in U-235, and the plutonium, for future use as fuel.   In fact, this choice does not occur in practice for the magnox fuel. Uranium metal and the magnesium alloy used as cladding for Magnox fuel are highly reactive and the spent fuel is not

suitable for prolonged storage under water or for
permanent disposal.   Reprocessing is therefore an
essential part of the magnox fuel cycle and seems likely
to remain so for the lifetime of the reactors, another ten
or fifteen years.   AGR fuel is enriched UO_2 clad in
stainless steel.   This fuel can be stored under water for
10-20 years, and longer in air-cooled stores.   PWR fuel,
UO_2 in Zircaloy, can be stored indefinitely under water.
PWR fuel, and possibly AGR fuel, can therefore be packaged
for direct disposal after a prolonged period - several
decades - of cooling in an artificially cooled store.
Plans for the direct disposal of spent PWR fuel have been
made in Sweden, Finland, Spain and, for civilian fuel, in
the USA.   The only large civilian plants for the
reprocessing of spent fuel outside the USSR exist in
Britain and France, though both India and Japan are
building their own.   Many other countries use the British
and French reprocessing plants.   BNFL have contracts
with eight countries for reprocessing oxide fuel in the
new Thorp plant.[8]

     The main objective in reprocessing fuel is to
conserve natural uranium.   The uranium recovered from
magnox fuel has been recycled through the enrichment plant
and used in AGRs.   Plutonium can also be used instead of
U-235 in thermal reactor fuel, as the so-called 'MOX'
fuel, about 3% PuO_2 in solid solution in UO_2.   MOX fuel
is currently being tested in Germany and in France, and
the uranium requirements can be reduced by up to 30% in
this way.   However, the most efficient way to use the
recovered plutonium is the use of a 'fast' reactor cycle,
in which the fast neutrons from the reactor core can be
used to 'breed' fresh plutonium in a blanket of U-238
surrounding the reactor core.   A typical fast reactor
cycle, such as is in use now at Dounreay and in the
commercial-size fast reactor in France, is an extremely
efficient fuel cycle.   The amount of fresh U-238
required is about 2% of that required for a thermal
reactor fuel cycle, and the depleted U-238 already in
store in this country would be a source of energy roughly
equivalent to our coal reserves if they were to be used in
this way.[9]

     Reprocessing is necessary in a fast reactor fuel
cycle because of the amount of fissile material in the
fuel.   Oxide fuel in a fast reactor contains about 25% of
PuO_2.   Reprocessing technology would therefore be an
essential part of an expanding role for nuclear power in

the 21st century, if this were envisaged, because of the
need by then to conserve the world's uranium reserves.

Uranium Ore to Fuel

Uranium is mined either by underground or open-pit
techniques, and the general impact of the operation is
similar to any other metal mining operation.   The annual
production now is about 40,000 t a year with a projected
rise to 100,000 t/y by the year 2000.   Since a typical
uranium content is 0.2%, the quantity of ore mined is
about one-fortieth of the quantity of coal required for
the same generation of electricity.   Uranium mining is
subject to the usual mining hazards but, because of the
smaller scale, the risks per unit of electricity generated
may be between 1/10 and 1/30 of those of coal mining.

An occupational hazard is the radiation and radon in
uranium mines, which has given rise to enhanced levels of
lung cancer in uranium miners in the past.   Ventilation
and other protective measures are now applied.   The
annual radiation doses to uranium miners in 61 mines in
the USA in 1979 and 1980 were in the range 18-29 mSv,
though UNSCEAR estimates that 10-12 mSv is a reasonable
annual average.[10]

The ore is processed by an acid or alkaline leaching
process at a location close to the mine, resulting in a
crude form of uranyl oxide known as "yellow cake".
Natural U-238 will be in secular equilibrium with the 14
members of the decay chain, and virtually all the members
of this chain below Th-230, including Ra-226 and Ra-222,
will be left behind in the mill tailings, together with
between 1 and 10% of the original uranium.   The main
environmental hazard then arises from the mill tailings
piles.   Tailings have tended to be kept in open,
uncontained piles or behind engineered dams with solid or
water cover.   In dry areas, radiation doses can be due to
radon emanation.   In wet areas, run-off water will
contain radium and may have to be treated before release
into water courses.   The radon release is a function of
the cover provided and it is likely that steps will be
taken to reduce emanation from abandoned mines, of which
there are some 1250 surface and 2000 underground in the
USA alone.   The collective dose due to emanations from
mill tailings may be about 6% of that due to the entire

fuel   cycle.[10]   The treatment of mill tailings is
compared to that of other wastes in a later section.

The further stages of purification, chemical
conversion and manufacture of fuel elements, including
enrichment in U-235 if necessary, involve low doses and
effluents, due to uranium and its short-lived daughters.

Reactor Operation

Nuclear reactors operate on essentially closed
cycles, but all give rise to some gaseous emissions and to
liquid effluent of low activity.   Magnox and AGR reactors
have graphite moderators and are cooled by CO_2 under
pressure.   The airborne discharge is due to the neutron
activation of impurities and to the loss of CO_2 containing
C-14.   The annual collective dose   from all UK power
stations has been assessed as 5 manSv, mainly due to C-14[3]
and the maximum individual dose from AGR stations is less
than 0.1 mSv/y.[11]   Radioactive liquid discharges arise
mainly from the storage ponds of spent fuel. Since most
of the reactors are on the coast, the resulting doses to
critical groups are low.   An exception is the station at
Trawsfynedd, where the discharge is to a lake.   The dose
to people eating fish from this lake was 0.35 mSv in
1982, reducing 0.25 mSv in 1987.[12]

Gaseous and liquid discharges from the world's
reactors are collected by UNSCEAR.[10]   In total, the
radiation doses from reactor operation are calculated to
account for most of the occupational exposures from the
nuclear fuel cycle and for 60% of the total exposures of
local and regional populations.   The majority of this
collective dose is due to emissions of C-14 and H-3.   The
individual doses, where they are calculated, are very low.

The Reprocessing of Spent Fuel

After a period of storage in cooling ponds at reactor
sites, spent fuel which is to be reprocessed has to be
transported to one of the three commercial reprocessing
plants, two of which are in France and one is at
Sellafield.   All packages or containers in which
radioactive material is carried have to be licensed in the
UK by the Department of Transport, and the regulations
conform to international standards set by the
International Atomic Energy Agency (IAEA) in Vienna.
Spent fuel has to be carried in the highest standard of

package, or "flask", which has to be shown to withstand
stringent accident conditions.[13] Some 7000 shipments of
spent fuel have been made in the UK without any release of
radioactivity. Transport by sea has also been carried
out for about 20 years, and the record has been good.[14]

At the reprocessing plant, the first task is to
separate the fuel from the fuel cladding and from other
parts of the fuel assembly. The fuel material is then
dissolved in hot nitric acid, when some gaseous fission
products are lost, and either trapped or dispersed to
atmosphere. The nitric acid solution is fed into a
solvent extraction plant, and the uranium and plutonium
taken into a solvent phase. The aqueous solution left
behind contains all but 0.5% of the fission products and
the higher elements of the actinide series, Am and Cm.
This solution constitutes the Highly Active Waste (HAL),
or Heat-Emitting Waste, because of the heat emission that
is due to the high specific activity. The uranium and
plutonium are purified in successive solvent extraction
cycles.[15] Final washings are stored and, at Sellafield,
discharged to sea after monitoring. The limits for
discharges both to atmosphere and to sea are set by the
Department of the Environment and the Ministry of
Agriculture, Fisheries and Food, acting together.

Reprocessing has been criticised on three grounds:
(i) the effluent is a major source of contamination; (ii)
reprocessing generates volumes of radioactive waste;
(iii) separating out plutonium adds to the danger of
proliferation of nuclear weapons. The industry has been
well aware of these criticisms and some comments are as
follows: (i) Discharges from Sellafield did increase in
the early 1970s due to a delay in reprocessing magnox
fuel, although they always remained below authorised
levels. In 1975, discharges from Sellafield constituted
the major source of collective dose in Europe. Since
then, many new treatment plants have been built; annual
discharges have reduced from 9000 TBq of β-γ emitters
in 1975 to below 100 TBq in 1988, and discharges of α-
emitters have fallen from 180 to 2 TBq.[16] The trend now
is to control and contain effluents, rather than to rely
on the dilute and disperse philosophy. (ii) Reprocessing
does not affect the quantity of radioactivity to be stored
and disposed of, but it does increase the volume of both
low-level and intermediate-level solid wastes. Low-level
wastes (LLW) are slightly contaminated rubbish which is
usually packed into drums and can be handled without

special precautions. Intermediate-level wastes (ILW)
require special handling techniques and some shielding,
but have a specific activity less than 1/1000 of that of
high-level wastes, so the heat evolution is not a
practical problem. Definitions of LLW and ILW in terms of
activity content have been published.[17] The volumes of
wastes arising over the whole lifetime of magnox and AGR
reactors are collected in Table 2,[18] from which it can
be seen that both the total volumes of waste and the
proportion of reprocessing wastes are smaller for AGRs
than for Magnox reactors. This is a direct result of the
higher fuel efficiency of the AGR, since much less fuel
has to be reprocessed for the same generation of
electricity. A change to PWRs would continue this trend.
Since magnox fuel has to be reprocessed, because of its
chemical reactivity, the reprocessing of advanced fuels in

Table 2   Lifetime Solid Waste Arisings (Packaged Volume (m^3) per
           GW(e))

	HLW	ILW		LLW	
		MAGNOX	AGR	MAGNOX	AGR
Reactor Operation		1,300	330	7,500	4,700
Fuel Reprocessing	130	21,250	2,600	200,000	25,000
Decommissioning		20,000	7,000	50,000	10,000
Total		42,550	9,930	257,000	39,700

Thorp will add less than 10% to the total volumes of
solid waste arisings to the year 2005.[8]   (iii) Guarantees
against illicit diversions of fissile material are given
in the Non-Proliferation Treaty and the monitoring regime
established both by the EEC and the IAEA.   However, it is
also true that the trend in civilian technology towards
more efficient, high burn-up fuel is leading to plutonium
which is not applicable to weapon design, because of the
low percentage of Pu-239 and the relatively high
percentage of higher isotopes.[16]   This additional safety
factor would apply also to fast reactor fuel cycles so
long as irradiated material from the blanket is
reprocessed together with material from the core.[9]

Table 3   Plutonium in Discharged Fuel

Reactor	Burn-Up MWD/tU	Pu-239	% at Discharge 240	241	242
Magnox	3,000	80	16.9	2.7	0.3
	5,000	68.5	25.0	5.3	1.2
AGR	18,000	53.7	30.8	9.9	5.0
PWR	33,000	56.2	23.6	13.8	4.9
FR (4th)		64	27	4	4

In many ways, therefore, the trend towards a high burn-up, efficient fuel will reduce the environmental impact of the fuel cycle in the future.

Disposal of Solid Wastes

Over 99% of the radioactivity in the entire fuel cycle will eventually appear in solid wastes of some form. Indeed, it is an important safety principle that all wastes stored as liquids or slurries should be converted to a solid form to reduce significantly any possibility of migration into the environment.

Low-level wastes can be treated to reduce the volume either by compaction or by incineration. Intermediate-level wastes are processed to minimise any release of activity by incorporation in cements, polymers or bitumen.[19] Decommissioning wastes are essentially the same as LLW, with some ILW. More than 95% of the radioactive content of the spent fuel appears in the aqueous phase of the first separation stage on reprocessing. This is stored in double-walled stainless steel tanks, cooled by pumping water through cooling coils. Safe storage depends on the maintenance of cooling; a HLW store involving acetates and nitrates caused the serious contamination accident in the USSR in 1954.[20] Solidification is therefore a most important safety step in the management of HLW.

Borosilicate glass is the solid form now used for HLW immobilisation on the commercial scale.[21] More advanced ceramic waste forms are also being developed.[22]

Vitrified HLW blocks have to be stored for some decades with artificial cooling, usually in a current of air, before they can be sent for final, permanent disposal.   Other categories of waste can be disposed of as soon as a repository is available.   In the UK, LLW is going to an upgraded Drigg site, and Nirex are examining sites at Sellafield and Dounreay for their suitability to host a deep depository for ILW and LLW.   Long-lived activity, as in HLW and much ILW, must be disposed of deep underground to increase the pathway back to man and to minimise the chance of unauthorised intrusion.

All countries with nuclear programmes have similar plans for permanent disposal of wastes, though some, such as France, Sweden and Germany, are more advanced than we are in Britain.   The rate of progress is dictated by the political difficulty of obtaining permission to investigate possible sites.[23]   Site investigation is essential, since the only credible mechanisms for radionuclides getting back to man is through leaching and migration in groundwater, and it is essential to measure the rates of groundwater flow through the rocks above a proposed repository.   Armed with that information, it is possible to evaluate the lifetime of the various barriers placed between the waste and man - (i) the durability of the waste form and its container;   (ii) the time for which the chemical conditions in a repository will remain alkaline and reducing, thereby inhibiting the migration of important radionuclides such as the actinides;   (iii) the time taken for any leached material to reach the surface; and (iv) the dilution effects in the biosphere.   The safety case is, of course, aided by the natural reduction of the radioactivity with time.[24]

A recent EEC study included dose estimations for HLW buried in named locations for which there was some geological and hydrological data in four different geological strata.[25]   Some results from this very comprehensive exercise may be noted.   The main result was that no significant dose arose in any of the locations studied until after 20,000 y, and maximum doses even at very long times were less than 0.1 mSv/y.   It seems, then, that the very stringent criteria set for waste repositories can be met;   the Department of the Environment has set a maximum risk as that corresponding to a dose of 0.1 mSv/y without time limit.[26]

In fact, the highest doses in the distant future are
likely to arise from the uranium mill tailings, not from
buried waste. Figure 2 is a plot of the total activity
against time of the inventory of ILW and LLW currently
assumed by Nirex to have accumulated by 2050,[27] together
with the activity of the HLW generated by a nuclear
programme of the same size and the activity of the amount
of uranium mill tailings needed for this programme.
After 10^4-10^5 y, the activity in the mill tailings is of
the same order as that of the ILW and LLW, and higher than
that of HLW, from which nearly all the uranium has been
extracted. But the mill tailings are stored close to the
earth's surface, while the ILW and HLW will be buried deep
underground. Therefore the individual and collective
doses from the mill tailings will be greater than those
from buried waste in the long term, though both will be
very low. Waste disposal will pose the lowest
environmental risk of any part of the nuclear fuel cycle,
and it is regrettable that the subject of waste disposal
has aroused so much unnecessary alarm.

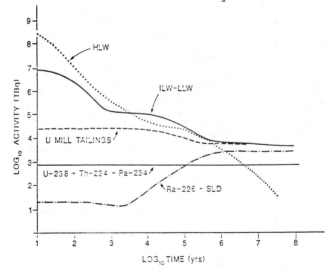

Figure 2   The total activity in various waste streams as a function
of time.
HLW - High level waste from 1025 GW(e)y of nuclear generation.
ILW + LLW - the contents of the planned Nirex repository (Rep. 71).
U-238 + Th-234 + Pa 234 - in ILW + LLW. from Rep. 71
Ra-226 + short-lived daughters - in ILW + LLW. from Rep. 71
U mill tailings - from 1025 Gw(e)y programme.

3   THE RISKS OF MAJOR ACCIDENTS

As well as considering the environmental impact of normal
operations, it is necessary to assess the risk of a major
spread of radioactivity arising from accidents. The two
places in the fuel cycle where large amounts of activity
are concentrated are the reprocessing plants and the
reactors themselves. The major hazard at reprocessing
plants would be associated with the spent fuel stores and
HLW storage tanks, both of which have to be kept cool.
However, it appears that there is enough time for remedial
measures to be taken if there is an interruption in the
supply of cooling water. Reactors also have to be
designed so that the fuel can be cooled at all times.

    Overheating of the fuel in a reactor can arise in two
ways (i) by a loss of control of the neutron
concentration and (ii) by a loss of cooling capacity,
which has to be maintained even after the reactor has been
shut down, because of the heat emitted due to the intense
radioactivity of the fission products. However, a
serious release of radioactivity will not occur even if
the fuel overheats unless the barriers due to the fuel
cladding, the primary cooling circuit ana the reactor
containment all fail at the same time.

    The principles on which safety instrumentation and
safety circuits should be designed so that faults arising
from component failure or external causes do not result in
a reactor accident are now well established.[28] Following
the pioneer work of Farmer and others in the 1960s and
1970s, methods of quantifying the risks of failure have
been developed to a high state of sophistication. A
quantitative risk analysis would be required by the
Nuclear Installations Inspectorate as part of the evidence
required before a reactor could be licensed. The
standards required were set out in a consultative paper.[29]

    The public inquiry into the proposal to build a PWR
at Sizewell included a long examination of the safety
case. The target set by the CEGB was to show that the
frequency of accident sequences that could lead to an
uncontrolled release were less than 10^{-7}/y, and that the
total of all such releases would be less than 10^{-6}/y.
This was the 'design basis' of the reactor. In addition,
the CEGB commissioned studies of "beyond the design basis"
accidents to ensure that no accident with large
consequences lay just beyond the design boundary.[30] The

risks were calculated as individual risks and as social risks, that is, the risk to the population of the UK of a spread of radioactivity due to an accident, calculated as the product of the frequency of an accident and the numbers affected.   The Inspector's final summary of social risk is reproduced in Table 4.[31]   In his report, the Inspector stressed that his conclusions on safety depended on his assessment of the overall strength of the safety case, on the competence of the operators and regulators and on the quality of operating and inspection procedures.   The quantitative analysis of risks was supplementary evidence that reinforced his conclusions.

Table 4   Estimated social risks to the UK population from Sizewell B

Cause	Annual social risk, deaths/y
Normal operations	16×10^{-4}
Design basis accidents	2×10^{-4}
Beyond design basis accidents	2×10^{-4}
Total	20×10^{-4}

That is, on average, 1 death in 500 years.

The value of a qualitative as well as quantitative audit of safety procedures became apparent after the first accident to a commercial reactor which resulted in extensive damage to the reactor core and consequent release of activity to the environment - the accident to a PWR at Three Mile Island, Pennsylvania, in March 1979. This was an example of a loss of cooling accident which was due to maloperation of the emergency cooling circuits after the reactor had been shut down.   The escape of radioactivity was actually slight because the containment remained intact.   The release is estimated to cause between 0 and 5 new cases of cancer in the area in the next 30 years,[32] but the public alarm was considerable. As a result of the investigation of the accident, two new institutes were set up in the USA to ensure that the lessons learnt would be applied, the Institute for Nuclear Power Operations and the Nuclear Safety Analysis Center.

Unfortunately these lessons were not heeded in the USSR.   The second severe accident to a commercial reactor, at Chernobyl in April 1986, was worse by many orders of magnitude.   The Russian RBMK reactors suffered

from a serious design fault, which could lead to loss of
control of the neutron levels if the reactors were run at
less than 20% of full power for long periods.   This
design fault was ignored by an operating team who were
attempting to carry out a test and this error was
exacerbated by the operators switching off or by-passing
emergency systems.[33]   The reactor experienced a power
surge which led to fragmentation of the fuel and to a
steam explosion which ruptured the containment.   The
resulting accident was comparable with the worst disasters
in the chemical industry, such as that at Bhopal.   Thirty
of the staff on the site died of burns and radiation
injury, and some 200 were treated for severe radiation
syndrome.   About 130,000 people were evacuated from the
neighbourhood and a long and expensive clean-up operation
was necessary.   The other three reactors on the site are
operating again, but the nearby village has not been
reoccupied.

     The design of the RBMK reactors would not have been
licensed in the UK; the Russians have admitted their
major faults and are taking steps to remedy them.[34]   The
radiological consequences of the spread of radioactivity
have been calculated by UNSCEAR.[10]   In the first year,
average doses due to Chernobyl reached 32% of background
in Eastern Europe, 11% of background in the USSR and 2-3%
of background in the UK.   They are falling with time.
The collective dose commitment to the world's population -
which will be received over 30 years - is about 6×10^5
manSv.   This may be compared with the annual background
of 12×10^6 manSv, and with the collective dose commitment
due to bomb tests in the atmosphere carried out between
1952 and 1981, which is 3×10^7 manSv.   It is highly
unlikely that the radiological impact of this accident
will be detectable in mortality statistics except,
perhaps, in the case of the workers on the site.   In most
cases, the extra doses received are less than the
variation in background dose between different areas of
one country.

     4   CONCLUSIONS

The conclusion to be drawn from this analysis is that the
radiological impact of all the normal operations of the
nuclear fuel cycle will be very low.   The relative
contributions in the very long term from the different
parts of the fuel cycle can be gauged from the collective
dose commitment, integrated to 10,000 years, and

normalised to 1 GWy of electricity generation, which was calculated by UNSCEAR;[10] this is reproduced in Table 5.

Table 5   Public Exposures from Solid Waste Disposal and Radionuclides in Effluents

Source	Collective Dose Commitment [manSv $(GWy)^{-1}$]
Mine and mill tailings (10^4 y)	150
Reactor operation	
LLW disposal	0.00005
ILW disposal	0.5
Reprocessing solid wastes	0.05
Globally dispersed radionuclides	63
Total (rounded)	200

The 'globally dispersed radionuclides' include the rare gases, Kv and Xe, and C-14 and this contribution could be reduced if required.   The dose from mill tailings should reduce as the industry moves to more efficient fuel cycles.

A significant increase will only occur if there are serious accidents, particularly to reactors.   The chances of this happening are low, but safety is only won by continued vigilance.   Following Chernobyl, the IAEA have strengthened their procedures and organisation aimed at promulgating uniform safety standards, and the world's nuclear operators have formed an international organisation (WANO) with the aim of exchanging information on any untoward incident and promoting the highest standards of operation and maintenance.   These initiatives deserve every support.

The advances made in the last 15 years in risk assessment applied to reactors have led to a better appreciation of the principles on which reactor safety should be based, and to the emergence of a series of designs of third-generation reactors which approximate to the ideal of being intrinsically safe, dependent for safe shutting down and cooling more on the laws of physics rather than on the reliability of engineered safety circuits.[28,35]   However, it will be a decade or more

before such new designs are commercially available, and
meanwhile efforts will continue to improve the safety
characteristics of the present designs.[36]   It is vitally
important to be able to demonstrate that everything
possible is being done to maintain and improve safety
standards.

REFERENCES

1. Health and Safety Executive.   'The tolerability of
    risk from nuclear power stations', H.M.S.O., London,
    1987.
2. National Radiological Protection Board, GS9.   Interim
    guidance on the implications of recent revisions of
    risk estimates and the ICRP 1987 Como Statement,
    H.M.S.O., London, 1987.
3. J.S. Hughes, K.B. Shaw and M.C. O'Riordan, NRPB-R227,
    H.M.S.O., London, 1989.
4. R.A. Clarke and T.R.E. Southwood, Nature, 1989, 338,
    197.
5. M.J. Gardner, A.J. Hall, M.P. Snee, S. Downes, C.A.
    Powell, J.D. Terrell, MRC Environmental Epidemiology
    Unit, University of Southampton, 5/2/90.
6. V. Beral, British Medical Journal, 1990, 300, 411.
7. S.C. Darby et al, British Medical Journal, 1985, 291,
    272.
8. Select Committee on the European Communities.
    Session 1987-88, 19th Report 'Radioactive Waste
    Management', H.M.S.O. London.
9. G.M. Jordan and L.E.J. Roberts, Phil. Trans. Roy.
    Soc. Lond., 1990, in press.
10. United Nations Scientific Committee on the Effects of
    Atomic Radiation 1988 Report, U.N. New York.
11. G.C. Dale (ed.) 'Safety of the AGR', 1982, CEGB and
    SSEB (London and Glasgow).
12. G.J. Hunt, 'Radioactivity in surface and coastal
    waters of the British Isles', MAFF Report No.19, 1988
    (Lowestoft).
13. J.D. Hart et al in 'Seminar on the resistance to
    impact of spent magnox fuel flasks', p.79, Institute
    of Mechanical Engineers, London, 1985.
14. A. Salmon, Nuclear Energy, 1984, 23, 237.
15. R.H. Allardice, D.W. Harris and H.L. Mills, Nuclear
    Power Technology, 1983, 1, 209 (Oxford University
    Press).
16. British Nuclear Fuels Ltd. evidence to Select
    Committee on the European Communities, 1988, see
    reference 8.
17. Radioactive Waste Management Advisory Committee,
    1984, Fifth Annual Report, H.M.S.O., London.

18. R. Flowers, Paper CEGB P21 to Sizewell B Power Station Public Inquiry, 1982. (CEGB, London).
19. R.H. Flowers, R.G. Owen in 'Radioactive Waste Management 2', 2, 63-75, Thomas Telford, London, 1989.
20. New Scientist, 124:3, 23/30 December 1989.
21. J.E. Mendel, Phil. Trans. R. Soc., 1986, A319, 49.
22. A.E. Ringwood, P.M. Kelly, Phil. Trans. R. Soc., 1986, A319, 63.
23. L.E.J. Roberts, Roy. Inst. Proc., 1988, 259.
24. R.H. Flowers, L.E.J. Roberts, B.J. Tymons, Phil. Trans. R. Soc. Lond., 1986, A319, 5.
25. PAGIS 'Performance Assessment of Geological Isolation Systems for Radioactive Waste Disposal Systems', 1988 (CEC, Luxembourg)
26. Department of the Environment, Disposal facilities on land for low- and intermediate-level radioactive wastes, H.M.S.O., London, 1984.
27. NIREX Report No.71, U.K. Nirex Ltd., Harwell, Oxon, 1989.
28. L.E.J. Roberts, P.S. Liss and P.A.H. Saunders, 'Power Generation and the Environment', Oxford University Press (in press), 1990.
29. Health and Safety Executive, 'The tolerability of risk from nuclear power stations', H.M.S.O., London, 1987.
30. F.P.O. Ashworth and D.J. Western, Nuclear Energy, 1987, 26, 233.
31. F. Layfield, 'Sizewell B Public Inquiry', Department of Energy Report, H.M.S.O., London, 1987.
32. J.G. Kemeny, Report on the President's Commission on the Accident at TMI, Washington D.C., 1979.
33. L.E.J. Roberts, Biologist, 1987, 34, 118.
34. J.H. Gittus et al, The Chernobyl accident and its consequences, UKAEA Report NOR 4200, 2nd ed., H.M.S.O., London, 1988.
35. M. Hayns, Atom, 1989, No. 392, 2.
36. A.R. Edwards, Nuclear Energy, 1985, 24, 241.

The Generation, Storage, Treatment, and Disposal of Nuclear Waste at Sellafield

L. F. Johnson* and W. F. Larkins
BNFL, WASTE MANAGEMENT UNIT, HEAD OFFICE, RISLEY, WARRINGTON
WA3 6AS, UK

1 INTRODUCTION

British Nuclear Fuels plc (BNFL) provide a complete nuclear fuel cycle service to the UK electricity generating industry and overseas customers. The Company's Head Office is located at Risley near Warrington with production sites located at Capenhurst near Chester, Springfields near Preston and Sellafield in West Cumbria. The Company also operates two Magnox nuclear power stations at Calder Hall (on the Sellafield Site) and Chapelcross in Dumfriesshire. The main services provided by BNFL are as follows:

i)   the purification and processing of uranium ore concentrates for UF_6 and UO_2 production (at Springfields);

ii) the enrichment of uranium (at Capenhurst);

iii) the manufacture of uranium and plutonium based fuels (at Springfields and Sellafield);

iv) the reprocessing of irradiated nuclear fuel (at Sellafield);

v)   the transportation of nuclear material.

As with many complex industrial processes, the above operations result in the inevitable formation of waste products. The purpose of this paper is to describe BNFL's strategies for the treatment, storage and disposal of the wastes generated by its operations.

Whilst wastes are generated at all the Company's sites, it is only at Sellafield that the full range arises. For simplicity, therefore, this paper concentrates on waste arisings at Sellafield.

2 WASTE CATEGORISATION

In its fifth annual report, the Radioactive Waste Management Advisory Committee defined three main categories of radioactive waste:

i)   low level waste;

ii) intermediate level waste;

iii) high level waste.

These are described in more detail below.

Low Level Waste (LLW)

Low level waste is classified as being that material containing radioactive substances, other than those very low level wastes which are acceptable for dustbin disposal (ie less that 400KBq in any 0.1m^3), not exceeding 4GBq/t alpha or 12GBq/t beta/gamma.

Solid LLW typically comprises materials from within controlled areas and the immediate working environment of personnel such that it has not been in direct contact with process materials. The physical nature of the waste is highly heterogeneous but includes the following:

i)   cellulosics: towels, cleaning tissues, floor
                   coverings, cardboard packaging and
                   wooden pallets;

ii)   plastics/rubbers: sheeting, bags, bottles, tubing,
                   protective clothing, footwear,
                   gloves and respirators;

iii) metals: electric cabling, scrap instruments,
                   engineering tools, scaffolding, buckets,
                   drums and containers;

iv)   miscellaneous: building rubble, excavation spoil
                   and glassware.

Liquid LLW or Low Active (LA) effluents arise primarily from pond storage of irradiated nuclear fuel and reprocessing operations. Other sources of arisings include waste treatment and decontamination operations.

Intermediate Level Waste (ILW)

Intermediate level wastes are classified as those wastes with radioactivity exceeding the threshold for low level wastes, but which do not require heat generation to be taken into account in the design of storage or disposal facilities.

There are in excess of 150 defined streams of intermediate level waste arising at Sellafield which fall into five main categories:

i)   cladding removed from fuel rods prior to/during reprocessing including magnox and aluminium swarf, stainless steel and zircaloy;

ii) slurries including sludges from corrosion of magnox fuel cladding, ion exchange resins from pond water treatment and ferric hydroxide flocs from LA effluent treatment;

iii) technological wastes comprising solid scrap items with either low alpha, high beta gamma or high alpha, low beta gamma contamination;

iv) ventilation extract filters;

v)   decommissioning waste.

High Level Waste (HLW)

High level waste is categorised as being that waste in which the temperature may rise significantly as a result of its radioactivity so that this factor has to be taken into account in designing storage or disposal facilities.

The only source of high level waste arisings within BNFL is the aqueous raffinate from the first solvent extraction cycle of reprocessing operations.

3 SOLID LOW LEVEL WASTE

In recent years, the average rate of raw (prior to any

treatment or packaging) solid LLW arisings from Sellafield has been about 25 000m^3/year although this is expected to rise to about 40 000m^3/year as new plants come on line. The waste is currently collected in purpose built skips which are transferred by rail to the Company's low level waste disposal site at Drigg which is located six kilometres south east of the Sellafield site.

At Drigg the waste is tipped from the skips into shallow trenches cut into an essentially continuous boulder clay stratum at 5-8 metres depth. The trenches are graded to allow trench leachate (resulting from rain and groundwater permeating through the waste) to be directed to the southern end where it is collected in a drainage system and discharged into a stream which runs to sea via a local river. As the waste level approaches the top of the trench it is covered with 1.5m of earth incorporating a geotextile sheet and small boulders to provide a stable surface from which tipping operations can continue. Six trenches have been filled in this way and a seventh is currently being filled.

In addition to LLW from Sellafield, the Drigg site also receives LLW from other BNFL sites, the United Kingdom Atomic Energy Authority, the Ministry of Defence, Nuclear Power Stations, Hospitals, Universities, radiochemical sites and various other industrial organisations. In total, waste from these sources currently combine to account for about 20% of total disposals.

In 1987, despite continuing evidence that trench disposal was radiologically acceptable, BNFL announced a major programme of improvements to operations at the Drigg site. This programme, which is aimed at conserving capacity within that area of the site that is currently consented for waste disposal and improving the visual impact of disposal operations, included:

i)   capping and provision of groundwater cut-off walls to limit rainwater infiltration and lateral migration of groundwater;

ii)   refurbishment of the existing leachate drainage system;

iii) containerisation of waste with compaction where appropriate;

iv)   orderly emplacement of containerised waste in
      engineered concrete vaults.

      A temporary cap has recently been installed over
the completed trenches. This comprises a 1:25 graded
earth mound incorporating a low density polyethylene
membrane. A permanent capping incorporating a thick
band of clay will eventually be installed once
settlement is judged to be complete. A groundwater
cut-off wall keyed into the underlaying clay has also
been installed to control a known pathway for leachate
migration.

      The refurbishment of the existing drainage system
has also been recently completed with the provision of
flow proportional sampling equipment. Work is now
underway to refurbish an existing marine outfall
(remaining from the sites previous use as a Royal
Ordnance Factory) in order to route leachate direct to
sea.

      Waste containerisation is being introduced on a
phased basis. Waste from non-Sellafield consignors is
now being routinely despatched to Drigg in either full
or half-height ISO freight containers. These are
emplaced in a new concrete vault (Vault 8) which was
introduced into service in August 1988. This has a
nominal capacity of about 180 000m^3 of containerised
waste with an estimated fill date of mid 1995.
Sellafield waste continues to be tumble tipped on a
temporary basis pending provision of a combined
compaction and packaging plant at Sellafield.

      Plans are being developed for the introduction of
compaction of both Sellafield and non-Sellafield waste.
It is currently envisaged that this will involve the
adoption of high force compaction of waste packaged in
either 2001 drums or nominal 1m^3 boxes.

      Consideration is also being given to a range of
possible future developments at Drigg. These include:

i)   alternative packaging concepts and, in particular,
      the introduction of grouting to minimise voidage
      thus reducing the extent of future site
      settlement;

ii)   the use of deeper vaults founded on engineered
      clay;

iii) alternative cap designs aimed at minimising the volume of earth fill needed to generate the cap profile and reducing the overall visual impact;

iv) alternative drainage concepts.

In addition to eliminating bulk voidage, grouting of the waste has a number of potential additional benefits:

i) the creation of a high pH environment which would be beneficial in terms of reducing the solubility of certain key nuclides;

ii) increasing the potential for re-sorption of nuclides;

iii) providing an additional barrier to waste/water contact.

4 INTERMEDIATE LEVEL WASTES

As noted earlier, the intermediate level waste category covers a wide range of wastes. The surface radiation dose from these wastes can vary from less than 0.01 to greater than 100Gy/hour, the heat outputs from less than 1 up to 500 watts/m^3, the pH of the aqueous streams from 1-14, and the physical size of the solid waste from a few millimetres to over 1m.

At present, these wastes are stored pending encapsulation in an inorganic cement matrix within specially constructed plants. Currently some 30 000m^3 of intermediate level waste is held awaiting conditioning in this way with further arisings estimated at about 3 000m^3 per year.

A major R&D programme started in 1982 to investigate the properties of a variety of encapsulation matrices in order to select appropriate formulations for each of the ILW streams at Sellafield. The programme was structured in four phases. Phase 1 included waste characterisation, specification of simulants for active waste and the identification of potential encapsulation matrices. Each of the preferred matrices was then assessed using small scale samples under phase 2 of the programme. Each matrix was evaluated for a number of properties including physical, thermal and radiation stability and mechanical strength. A multi-attribute decision analysis technique was then used to select the preferred matrix for each ILW stream for more detailed study in phase 3. The results from these studies have shown that inorganic cements are the most suitable encapsulant for all the ILW streams examined. For most ILW it has been possible to adopt one of two cement formulations as the preferred matrix. These are blends of either ground granulated blast furnace slag (BFS) or pulverised fuel ash (PFA) with Ordinary Portland Cement (OPC). The final phase of the programme was aimed at evaluating the sensitivity of the product performance to variations in process parameters.

A number of generic studies which were particularly relevant to disposal of the waste have also been carried out. These include:

i)   the effects of chemical, microbiological and radiological degradation of the waste which might enhance the solubility of long lived isotopes;

ii)   the effect of sorption of the long lived isotopes by the matrix;

iii) the effect of organic components within the waste on the subsequent mobility of activity away from a final repository.

Three main plants for the encapsulation of intermediate level waste are now at various stages of completion at Sellafield:

i)   Encapsulation Plant 1 (EP1) which will process magnox swarf (magnesium/aluminium alloy fuel cladding). This plant is due to commence active operation shortly;

ii)   Encapsulation Plant 2 (EP2) which will process stainless steel fuel hulls and slurries to be produced when oxide reprocessing commences in 1992;

iii) the Waste Packaging and Encapsulation Plant (WPEP) which will process flocs and maintenance wastes from the liquid effluent treatment plants. This plant is scheduled for completion during 1992;

A purpose built plant has also been provided for processing plutonium contaminated materials. Completion and operation of this plant awaits regulatory approval of the final disposal form.

All the above plants are designed to encapsulate waste in stainless steel 500l drums to a design which has been agreed with UK Nirex Limited. Two basic processes are involved. Solid wastes are to be in-fill grouted whilst sludges/slurries will have the dried grout components added (the water being already present in the waste) with in-drum mixing via a "lost paddle". In all cases the waste is cured for 24 hours in a curing cell prior to adding a layer of clean capping grout (to seal any surface activity) and a lid.

An additional facility, the Encapsulation Pretreatment Plant (EPP), is also proposed for the late 1990's. The function of this plant will be to package waste which has been retrieved from storage prior to encapsulation in EP1 or EP2.

The products from the encapsulation plants will be transferred to purpose built stores where they will be held until a permanent deep repository is available. Geological investigations are currently underway at both Dounreay, near Thurso in Scotland, and at Sellafield to determine the suitability of the sites for construction of a deep repository.

5 HIGH LEVEL WASTE

High level waste is the concentrated aqueous raffinate from the first solvent extraction stage of fuel reprocessing. The reprocessing operation is optimised to retain about 97% of the radioactivity associated with fission products in this waste stream thus minimising activity appearing in the intermediate and low level waste streams. The constituents of the HLW liquor are variable depending upon the "burn-up" of the fuel reprocessed, process additives and the extent of fuel corrosion during storage pending reprocessing. The principal constituents are, however, as follows:

i)   fission products which could include tritium and any element in the periodic table from germanium to actinium;

ii)   transuranic elements including neptunium, americium and curium, the yields of these elements increasing as the fuel "burn-up" increases;

iii) unextracted uranium and plutonium;

iv)   fuel alloying agents such as iron, aluminium,
       silicon and molybdenum;

v)   elements such as iron, chromium and nickel from
       minor corrosion of the stainless steel process
       pipework;

vi)   process liquors and additives comprising of nitric
       acid, kerosene/tributylphosphate (present as an
       emulsion) gadolinium and boron (added for
       criticality control) and organic degradation
       products such as dibutylphosphoric acid and
       monobutylphosphoric acid.

Annual arisings of HLW total some $100m^3$. This is
currently concentrated by reduced pressure evaporation
for storage in stainless steel tanks. The tanks are
heavily shielded, double contained with multiple
cooling and agitation systems. Empty tanks are
available as a standby. About 1 300m^3 of HLW is
currently held in store in this way.

Current UK policy relating to high level wastes
involves surface storage for at least 50 years prior to
disposal in order to allow them to 'cool'. Although
liquid HLW storage is accepted internationally as a
safe system, it is not considered to be a long term
solution. Plans are now well developed to process all
HLW (both future arisings and the backlog in storage)
into an intrinsically safer vitrified form that would
be suitable for eventual disposal. This involves
calcination of the HLW concentrate followed by
incorporation into a borosilicate glass which is
enclosed in stainless steel containers of 150l nominal
capacity. This product was chosen for its stability,
radiation resistance, thermal properties, mechanical
strength, chemical durability and leach resistance.
Processing in this manner also has the added advantage
of reducing the volume of HLW for storage by between
2:1 and 3:1.

The glass formulation, which was selected after
extensive studies, has an amorphous three dimensional
network of formers such as SiO_2, B_2O_3 and P_2O_5
covalently bonded together with oxygen "bridges".[5] The
majority of the elements in the high level waste
inventory act as network modifiers when introduced into
the glass thus breaking the covalent oxygen "bridges"

to form ionic, non-bridging oxygen bonds.

This process will be carried out in the Windscale Vitrification Plant (WVP) which is currently undergoing inactive commissioning and will commence active commissioning by mid 1990. The process initially involves feeding pre-prepared $10m^3$ batches of HLW liquid into an inclined rotary calciner, the resulting off-gas being extensively scrubbed before discharge. The output from the calciner will then be mixed with glass frit in an induction melting furnace at $1150^{\circ}C$, the glass product flowing under gravity into a stainless steel container held at $500^{\circ}C$ to reduce shock and ensure successive pours adhere together. A lid will then be loosely fitted and the container and lid completely sealed by an automatic fusion welding technique. The sealed containers will then be externally decontaminated and transferred to an adjacent air cooled product store.

6 LIQUID EFFLUENTS

Active liquors are generated by a wide range of operations but the principal sources of arisings are pond storage of irradiated nuclear fuel awaiting reprocessing and solvent washing stages downstream of the first solvent extraction cycle of reprocessing operations. The wastes consist mainly of water, together with small amounts of other material some of which are radioactive. These materials are in one of three forms:

i)   soluble materials which can only be separated by chemical treatment;

ii)   insoluble particles which can be separated by physical methods;

iii)   traces of organic solvents which can also be separated by physical methods.

Most of the fuel handled at Sellafield to date has been Magnox fuel which is clad in a magnesium aluminium alloy. In order to minimise the corrosion of this fuel during storage in water filled ponds, a number of measures have been adopted including:

i)   dosing storage water to pH 11.5 with sodium hydroxide;

ii) controlling levels of chloride, sulphate and
silicate to as low a level as is reasonably
practicable;

iii) maintaining the pond water temperature in the
range 10-15°C.

Such corrosion as does take place results in
radioactive caesium and strontium being released to the
pond water. In order to maintain the levels of both
radioactive and non-radioactive ions at acceptable
operational levels the pond water is purged and
replaced with demineralised water, dosed to the correct
pH. This amounts to a total of about 3 000m³ of purge
per day for all the Sellafield ponds. The pond purge
is treated in the Site Ion Exchange Effluent Plant
(SIXEP) prior to discharge to sea. SIXEP uses simple
filtration and ion exchange to treat the pond water.
Initially the purge water is passed through a sand
filter to remove magnesium (which competes with caesium
and strontium during the ion exchange stage) and
suspended solids (which would otherwise blind the ion
exchange columns). The liquor is then conditioned to a
pH of about eight in a carbonating tower, this being
the optimum pH for the subsequent ion exchange stage.

An inorganic ion exchanger, Clinoptilolite (a
natural zeolite), was selected following a
comprehensive evaluation programme. This material has
a high selectivity for caesium and strontium and is
compatible with subsequent encapsulation in a cement
based matrix for eventual disposal.

In terms of performance, the residual solid level
in the discharge from the plant is less than 0.5ppm and
caesium decontamination factors of approximately 1 000
have been achieved.

Medium active liquors arising from solvent washing
cycles are sent to a new evaporation plant. The
effluent is first passed through a caustic scrubber to
remove radioactive iodine which would otherwise be
vented to atmosphere. Free phase solvent entrained in
the liquor is removed by weirs in the feed stock
storage tanks. The liquor is then transferred to a
stirred, steam heated conditioning vessel where it is
acidified to pH2 and heated to 90°C. The conditioned
feed is then stripped of residual solvent, cooled and
passed to a vacuum evaporator for concentration at
60°C. The resulting liquor is then despatched to a

storage facility for decay storage pending further future treatment.

The stored evaporator concentrate (about 1 000m^3/year) will eventually be treated in a new plant, the Enhanced Actinide Removal Plant (EARP) along with certain bulk low active effluents (about 250m^3/day). The acidic feed liquor will be neutralised with sodium hydroxide which, with iron levels well in excess of 100ppm, will result in formation of a ferric floc. The bulk alpha activity associated with the effluent streams will be co-precipitated with this floc. Floc/liquor separation and floc concentration will then be achieved by ultrafiltration and the resultant supernate will be discharged and the floc encapsulated in a cement based matrix pending future disposal. It is currently expected that decontamination factors of up to 100 will be achieved on the bulk effluents and up to 500 on the concentrates.

The remaining low active effluents (up to 5 000m^3/day), which do not merit treatment in EARP, will be neutralised in a new plant (which will replace an existing facility) and discharged.

The other principal liquid arising is spent solvent from reprocessing operations. The solvent is a mixture of tributylphosphate and odourless kerosene (TBP/OK). The spent solvent is currently stored but treatment is eventually envisaged particularly since the rate of arisings is expected to rise significantly when the new Thermal Oxide Reprocessing Plant is brought into operation. The precise method of treatment has yet to be confirmed but the current preference is for the adoption of alkaline hydrolysis. The TBP/OK mixture would be treated with sodium hydroxide at about 130°C. The liquor would then be allowed to settle, forming three layers:

i)   an upper layer of kerosene which would be incinerated;

ii)   a middle layer of sodium dibutylphosphate which will be treated with hydrogen peroxide and neutralised with nitric acid prior to discharge (via EARP if the activity level were found to be above specification);

iii) a lower caustic layer containing the bulk of the

activity which would be pre-treated and fed to
EARP.

7 GASEOUS WASTE

There are three main sources of contaminated gaseous
arisings:

i)   building ventilation air;

ii)   active containment extract;

iii) process vessel off-gases.

The principal contaminants include volatile
radionuclides such as I-129, Kr-85, Ru-106 and noxious
gases such as NO_x and SO_x.

BNFL use standard industrial process gas treatment
systems, often in combination, to remove the activity
from these streams before discharge. The range of
cleaning systems used include electrostatic
precipitators, wet scrubbers, chemical clean-up
systems, packed beds and high efficiency particulate
air (HEPA) filters.

The treatment systems, whilst being standard
processes, are modified to meet the needs of active
plant operation. This usually involves modification of
designs to allow remote maintenance in order to reduce
operator dose uptake.

8 CONCLUSION

The generation, handling, treatment and disposal of
radioactive waste within BNFL has been reviewed.
Whilst BNFL has a comprehensive policy of waste
treatment, the Company is also sensitive to public
concerns on the nuclear waste issue and therefore
adopts, as far as practicable, processes and operating
practices that minimise the extent of waste arisings.

Severe Accidents

P. N. Clough
AEA TECHNOLOGY, WIGSHAW LANE, CULCHETH, WARRINGTON WA3 4NE, UK

1 INTRODUCTION

Although the nuclear power industry as a whole has an
excellent safety record and an extremely thorough approach to safety, public perception has been conditioned by a small number of major reactor accidents.
The accidents at Windscale in 1957, Three Mile Island
Unit 2 (TMI) in 1979, and Chernobyl Unit 4 in 1986,
have all left a deep public impression. The Chernobyl
accident in particular has provided a vivid example of
the very adverse consequences for human health and the
environment of the worst type of nuclear reactor accident. However, it is misleading to associate the Chernobyl accident with the earlier events at Windscale and
TMI. The Windscale accident occurred not in a power
reactor, but in a production pile at an exploratory
stage in the application of nuclear fission. Many
safety lessons were learned from Windscale which
benefited the subsequent civil nuclear power programme.
TMI was a pressurised water reactor (PWR) of a type
common in the West, whereas Chernobyl was a channel-
type boiling water reactor (BWR) of the RBMK design,
peculiar to the Soviet Union. The approaches to design
safety and built-in safety systems for these two reactors were very different. As will be described, the
health and environmental consequences of the TMI accident were negligible, and the systems designed to
protect the public and the environment were ultimately
successful there. In this respect, TMI was not a
severe accident. However, the TMI accident led to a
resurgence of interest in the chemistry of severe accidents generally, and a recognition that this aspect had
been neglected compared with the physics. Chernobyl

was a severe accident by any standards, and must be
placed in a unique category.

The U S Reactor Safety Study[1] (RSS) provided the
first full systematic analysis of the probabilities and
consequences of severe accidents in commercial nuclear
power plant. Its chief purpose was to determine the
public risk associated with nuclear power development,
and it set a pattern for all subsequent risk studies,
including the recent very exhaustive analysis in the
USNRC document NUREG-1150[2]. Preventing accidents is a
first concern of the nuclear power industry, and a
major part of such risk studies is devoted to identifying which combinations of plant and safety system
failures might lead to severe accidents, and the associated probabilities. Brief consideration is given
to accident prevention and probabilities in the following section. Of more interest to the chemist are the
succeeding stages of risk analysis. The second stage is
concerned with determining the amount and nature of the
radioactive material which might escape from a nuclear
plant in a severe accident, that is, the source terms.
The health and environmental consequences of such
releases are then analysed in the third stage. Some of
the more interesting chemical aspects of radionuclide
behaviour identified with these stages are discussed in
sections 3 and 4 respectively. The relevant evidence
from TMI and Chernobyl is also examined.

2 NATURE AND PROBABILITIES OF SEVERE ACCIDENTS

Safety is a key feature from the earliest conception
and design stages of nuclear power plants. A
philosophy of defence-in-depth has been followed in all
Western designs. Thus, a combination of inherent
safety features, and passive and active engineered
safety systems, is present in these designs. In normal
operation, and in a wide range of off-normal conditions, the highly radioactive fission products will
remain safely locked within the fuel in the reactor
core. The fuel represents one of a series of barriers
which must be breached in an accident if there is to be
a release of radioactivity to the environment (Figure
1). Modern reactors employ UO_2 or mixed U/PuO_2 as
fuel. These are ceramic materials, with melting points
around 2800 $^\circ$ C, and the fuel pellets are clad in a
metallic sheath. The fuel must overheat by many
hundreds of degrees above its normal operating temperature for the conditions characteristic of a severe accident to develop. The cladding will then rupture,

and certain fission products will start to diffuse quite rapidly out of the oxide matrix. A basic feature of severe accidents is that some unlikely coincidence of system failures must arise to lead to gross over-heating of the fuel. However, this condition alone will not necessarily result in a large release to the environment. Further barriers must be penetrated first, as shown in Figure 1. These are discussed in section 3.

Fuel overheating may arise in two main types of accident. Events such as failure of power supplies can induce transients in the reactor cooling, in which case diverse systems are provided to shut the reactor down. If these all fail, there is a possibility that under-cooling of the core might occur. Very rapid fuel over-heating could result, attaining fuel melting tempera-tures on a timescale of seconds. An accident of this type is essentially ruled out in PWRs, since the over-heating generates voiding of the water moderator in the core, and the reactor shuts itself down. This is an important inherent safety feature. The second type of accident is generally referred to as a loss-of-cooling accident, or LOCA. This is initiated by some failure

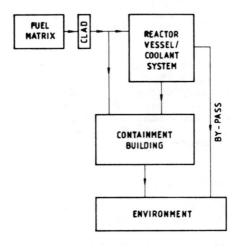

**FIG. 1 SCHEMATIC OF THE BARRIERS TO RELEASE
AND TRANSPORT PATHS FOR RADIONUCLIDES
IMPORTANT FOR SOURCE TERM EVALUATION**

of the coolant circuit, resulting in inadequate heat
removal from the core. The reactor is immediately shut
down, and fission ceases, but the decay of fission
products in the core still represents a large heat
source. Reactors are provided with multiple emergency
cooling systems to control this situation, and only if
these all fail will a severe accident develop. Inadequate cooling then results in heat-up of the core, and
damaging fuel temperature may be reached on a timescale
of between 30 minutes and many hours, depending on the
reactor type and the particular accident sequence.

The methods developed in the RSS[1] for determining
which combinations of system failures could lead to
melting in a reactor core involve a very detailed
analysis of the plant design, coupled with the application of logical analysis methods such as fault and
event trees. These methods can be used to calculate
the frequencies of various representative types of
severe accident for a specific reactor. A distinction
must be made between the frequency of core melt itself,
and the frequency of a consequential large release of
radioactivity to the environment. In the case of a
PWR, for example, the reactor coolant circuit and the
strong containment building must also be breached.
Analysis shows that this will occur in only a small
proportion of core melt accidents. Table 1 gives the
calculated total core melt frequencies, and frequencies
for large activity release, for the U S Surry 1 and U K
Sizewell B PWRs.

The lower frequencies for Sizewell B reflect a number
of significant safety improvements in a current design
over the 1960's Surry design. These values illustrate
the extremely low probabilities of severe accidents for
Western PWRs, and the results from analyses of many
other nuclear plants in the Western world typically
fall within this range.

What, then, happened at TMI and Chernobyl ? TMI Unit 2
was a Babcocks and Wilcox 900 MWe PWR. On the 28

Table 1 Severe Accident Frequencies predicted for two
        PWRs (per reactor year)

	Core melt	Large release	Reference
Surry 1	4.1(-5)	5.2(-6)	2
Sizewell	1.3(-6)	4.4(-8)	3

March, 1979, a fault condition developed which rapidly
evolved into a type of LOCA. Coolant was being lost
from the core cooling circuit, and in response, the
emergency cooling system started up automatically. Had
the system been left to itself, a safe shut-down would
have followed. However, the operators misunderstood
the nature of the fault, and turned off the emergency
cooling. As a result, after a few hours the core began
to boil dry, and overheated, some regions becoming molten. At this stage, the operators realised their mistake, and reinstituted the emergency core cooling.
Fortunately, the core cooled down fairly rapidly, and
the accident was brought under control. The reactor
vessel and containment building remained intact. Although a large amount of radioactive material was
transported from the core into the containment, very little   escaped to the environment. The main feature of
TMI is that, inspite of human error, the designed-in
safety ultimately succeeded.

      The Chernobyl reactor was not, of course, a
Western design. The arrangement of an RBMK core is
shown schematically in Figure 2. The fuel pins are
situated in vertical zirconium alloy pressure tubes.
High pressure cooling water flows upwards through these
tubes, and boils. The steam produced drives the turbines. The core contains some 1600 such pressure tubes
which pass through the massive moderator block constructed of 2000 tons of graphite. This design, as
implemented at Chernobyl, had inherent safety
weaknesses which made it unacceptable by Western safety
standards[4], and which played a key part in the accident. Somewhat ironically, the accident occurred
during a safety experiment. Erratic behaviour on the
part of the operators, which included the over-riding
of all built-in safety systems, brought the reactor to
an unstable low power condition on the 26 April, 1986.
A transient event then caused a runaway of the nuclear
fission reaction, generating a prompt and massive
energy spike. The reactor core and building were blown
open, allowing a direct release of radioactivity to the
environment. This catastrophe, which demonstrated the
international dimension of severe nuclear accidents,
met with a massive international response. Much effort
has been devoted worldwide to analysing the causes of
the Chernobyl accident, and ensuring that all possible
safety lessons have been learned[4-6]. The Soviet Union
has made major safety improvements to the RBMK reactors
still in operation[6].

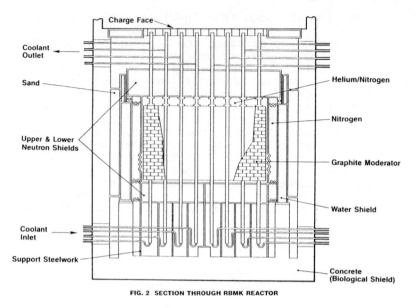

FIG. 2 SECTION THROUGH RBMK REACTOR

3   SOURCE TERMS

The fission of U235 in a thermal reactor generates some
30 lighter elements in significant yields, split into
two groups centred around mass numbers 95 and 140.   In
addition, neutron capture processes result in the
production of some half-dozen important transuranics.
Several hundred different isotopes are present immediately after a reactor is shut down, representing a
core inventory of several thousand million Curies (MCi)
for a typical 1000 MWe commercial power reactor.   Some
simplifying approach to this chemical and physical
diversity is necessary in analysing severe accident
source terms.   The RSS established the principles of a
simplification scheme which has stood the test of time
well.   Many isotopes have such small yields or such
short half-lives that they will not contribute significantly to accident consequences, and can be
eliminated.   Of the remainder, the chief selection
criteria are uptake by humans and the resulting health
effects.   In this way, a short-list spanning 54
isotopes of 26 elements was drawn up.   Although the emphasis in the RSS was on human health, concern about
land contamination and dose via the food chain ensured
that the list includes isotopes of Cs, Ru, Sr, Ce, and

the lanthanides and actinides, which are the most important for long-term environmental damage. The Chernobyl accident has confirmed the validity of the selection process in the RSS.

The RSS adopted a broad brush approach to
chemistry which persisted through many subsequent risk
analyses for light water reactors (LWR)[2]. The 26 elements were grouped into 7 radionuclide groups based on
chemical and physical similarity. Within a group, the
fraction of the initial core inventory of each member
released to the environment in a severe accident was
taken to be the same. This use of release fractions,
rather than actual activity releases for individual
isotopes, is an established convention in source term
analysis. Each group is represented by a lead member.
Thus, Cs covers all the relevant isotopes of caesium
and rubidium, whilst La covers 12 transition, rare
earth and actinide isotopes. Some important
radionuclides included in the groups, and their calculated inventories in the Chernobyl reactor core at the
time of the accident[4], are shown in Table 2. Low-
active and stable isotopes must also be considered for
some elements, since these can contribute most of the
mass inventory which is important in determining chemical interactions. Table 3 shows the estimated source
term release fractions for each group in the Chernobyl
and TMI accidents[7]. The contrast in the releases for
the two accidents clearly illustrates how different
they were.

In analysing source terms as part of a risk study,
it is convenient to divide the accident into a number
of distinct release and transport stages. These corre-
late with the barriers to release shown in Figure 1.
For a large release to occur, all the barriers must be
breached, and the fission products must exist in mobile
forms. The important mobile forms are gases and
vapours, aerosols, and to a lesser extent, solutions.
Source term analysis is mainly concerned with estab-
lishing what fractions of each radionuclide will be in
airborne form, and thus available for release to the
environment, when the final barrier fails. Chemistry
plays the key role in determining these forms. Most
modern thermal and fast reactors use oxide fuel, and
there is much common ground in fission product release
from fuel in severe accidents for LWR, sodium-cooled
fast reactors (FR), and advanced gas-cooled reactors
(AGR). Once into the reactor coolant system, chemical
interactions with the coolant become important, and be-

Table 2   Radionuclide Groups and Inventories in the
           Chernobyl Core

Group	Nuclides	$t_{\frac{1}{2}}$	Chernobyl Inventory MCi
Xe	Xe133	5.2d	178
I	I131	8.04d	80
	I133	20.8h	138
Cs	Cs134	2.06y	3.0
	Cs137	30.0y	6.4
Te	Te132	3.2d	110
Ba	Sr89	50.5d	99
	Sr90	29.1y	5.4
	Ba140	12.7d	156
Ru	Mo99	66h	149
	Ru103	39.4d	115
	Ru106	368d	24
	Rh105	1.5d	69
La	Y91	58.6d	126
	Zr95	65.5d	158
	Nb95	35.1d	159
	La140	40.3d	161
	Ce141	32.5d	151
	Ce143	33.0d	119
	Ce144	285d	105
	Pr143	13.6d	142
	Nd147	11.0d	58
	Np239	2.36d	1278
	Pu241	14.4y	2.7
	Cm242	153d	0.3

Table 3   Source Terms for TMI and Chernobyl

	\multicolumn{7}{c}{Fraction of Core Inventory Released}						
	Xe	I	Cs	Te	Ba	Ru	La
Chernobyl	1.0	0.4	0.25	>0.1	4(-2)	5(-2)	3(-2)
TMI-2	<8(-2)	2(-7)	0	0	0	0	0

Note: $3(-3) = 3 \times 10^{-3}$

haviour of individual fission products is specific according to reactor type. Analysis of LWR severe accidents has been much more extensive than for other reactor types. For this reason, and because the TMI and Chernobyl accidents both involved water-cooled reactors, the remainder of this section concentrates on the chemical aspects of LWR severe accidents. The discussion focuses on iodine, caesium and tellurium, the so-called volatile fission products, and on ruthenium as an example of a non-volatile fission product. The isotopes I131 and Te132 dominate the early health effects for a severe accident, whilst Cs134, Cs137 and Ru106 play major roles in long-term environmental contamination.

Release from Fuel

<u>Solid Fuel</u> The atoms formed by the neutron reactions initially exist as a dilute solution in the uranium dioxide fuel matrix, the atom fraction of any product element being well below 1% even at the highest fuel burn-ups currently employed. Some fission products, such as the lanthanides, form true solid solutions with UO_2, and show no tendency to migrate. Others, notably iodine, caesium and tellurium show little chemical interaction with the oxide lattice[8], and diffuse through the solid along thermal gradients. Caesium will form uranates under certain conditions, but the oxygen potential is too low to stabilise these in fuel under normal operation. I, Cs and Te show some tendency to diffuse out of the fuel even in normal operation, and small fractions (< 1%) of their inventories accumulate in the fuel-clad gap as burn-up progresses. Ruthenium shows little tendency to diffuse from the fuel at normal operating temperatures (ca 800 $^{\circ}$C), although highly-rated fuel develops separate noble metal phases at high burn-up.

Oxide fuel consists of UO_2 grains a few tens of micrometres in diameter, compressed into cylindrical pellets, typically about 0.01 metres in diameter. In a severe accident, the fuel cladding will rupture quite early. High temperatures and the steep temperature gradients promote the diffusion of the mobile elements out of the fuel grains. Vapour and surface diffusion processes then transport them to the fuel pellet surface, where they may be released into the gas phase. The core of an LWR will boil dry in the early stages of a severe accident, so that the ambient medium will be a mixture of high temperature steam and hydrogen. The

hydrogen is the product of the reaction of steam with
the zirconium alloy fuel cladding employed in LWRs:

$$Zr + 2H_2O = ZrO_2 + 2H_2 \qquad (1)$$

This highly exoergic reaction (ΔH^O = -586 kJ/mol)
proceeds rapidly at temperatures above 1200 ^{O}C, and is
a major contributor to the heat sources driving up core
temperatures, in addition to the decay heat. The
presence of the hydrogen formed by the reaction ensures
that chemically reducing conditions will prevail in the
core and reactor coolant system throughout most of an
accident. Soviet experts believe that this reaction
played a key part in the initial temperature excursion
in the Chernobyl accident, when fuel temperatures rose
to around 2700 ^{O}C in a few seconds. The reaction was
also responsible for generating a large amount of
hydrogen in the TMI accident, which led to a hydrogen
deflagration in the containment building[9].

     Much experimental evidence have been accumulated
on the release rates of fission products from oxide
fuel for simulated LWR severe accident conditions[7].
This shows that Kr, Xe, I, Cs and Te are released much
more readily than any of the other fission products,
and that their release will be almost total after a few
minutes at temperatures above 2000 ^{O}C. The stage at
which chemical reaction of fission products with each
other begins is still somewhat uncertain. Such reactions might be expected to occur first at fuel grain
boundaries, as fission products accumulate after escap-
ing from the grains. Thus, caesium and iodine may
react together within the fuel to form CsI, which is
thermodynamically favoured over a range of reducing
conditions. Alternatively, if Cs and I are released
separately but simultaneously from the fuel, CsI vapour
may be formed by a gas-phase reaction. Again, thermodynamic calculations predict CsI vapour as the major
iodine species in the gas phase, and reaction kinetic
modelling suggests that it will be formed very rapidly
under accident conditions[10]. The fission yield of Cs
is more than 10 times that of I, so that CsI could account for all of the iodine. Whatever the sequence of
the chemical and physical processes, there is now much
experimental evidence that CsI is the dominant form of
iodine released from fuel in a reducing atmosphere.
The TMI accident provided the stimulus for this renewed
examination of iodine chemistry[11, 12], and for investigations of a wide range of other severe accident
chemistry issues during the last decade[13].

It is now well accepted that CsOH will be the predominant form of the excess caesium released from fuel into steam/hydrogen atmospheres. Tellurium shows more complicated behaviour, since it reacts readily with the Zircaloy cladding material of LWR fuel. There is some evidence that the minor tin component in the alloy plays a key role through formation of SnTe. Oxidation of the Zircaloy by steam results in the release of Te. Thus, although elemental Te is released initially from the oxide pellets at a similar rate to I or Cs, the reaction with Zircaloy holds up the further transport of this element into the coolant channels. The release of Te will be delayed in time, and may be less complete, than that of I and Cs.

Molten Fuel Fuel liquefaction will occur during the final stages of core heat-up in the reactor vessel. In an LWR accident, this may occur well below the melting point of UO_2, since interaction with molten Zircaloy cladding can produce mixed phases with melting points as low as 2200 $^{\circ}$C. Liquefaction will lead to a loss of core geometry and slumping of the core debris. Melt-through of the steel reactor vessel will ultimately ensue. A further important stage for the release of fission products can then arise when the molten core debris contacts and attacks the concrete base of the reactor building. Gas bubbles released from the decomposing concrete sparge through the melt and promote the release of vaporised radionuclides. As the bubbles break the melt surface, the vapours rapidly cool and condense to form aerosols which are released to the containment atmosphere. This so-called molten core-concrete interaction (MCCI) may be effective in releasing those fission products which are immobile in solid UO_2, and so were not released earlier. The chemistry of MCCI is extremely complex, and is still the subject of detailed experimental investigation[13]. The elemental forms or oxides of U, Zr, Fe, Cr, Ni, Si, Al, Ca and Mg, as well as of the radionuclides, may be present in the melt, in two or more liquid phases. The nature of the concrete plays a key part, limestone concrete releasing more CO_2 than basaltic, and hence being more effective in promoting radionuclide release. The role of zirconium is also very important. If unoxidised Zr is present, the oxygen potential will be low. This will favour the release of fission products with volatile lower oxide forms, such as LaO. Once all of the Zr has been oxidised by CO_2 and H_2O from the concrete, the oxygen potential will rise steeply. Release of radionuclides such as ruthenium with

volatile oxides (RuO_3, RuO_4) will then be favoured.
Models designed to predict radionuclide release in MCCI
are based on equilibrium chemical thermodynamics. A
difficult problem in such models is the representation
of the activity coefficients of melt constituents,
since there is evidence of large negative deviations
from ideal behaviour for several important species.

Transport in Reactor Coolant System and Containment

Radionuclides released from the core during over-
heating in the reactor vessel could, in some accidents,
be substantially retained in the coolant system. Temperature, chemical behaviour, and the length and
geometry of the flow path from the core to the circuit
break are key factors in determining the extent of such
retention. Figure 3 shows schematically the important
chemical and physical processes in the coolant system.
For LWR accidents, the chemical interactions of a fission product can be broadly categorised into reactions
a) with the steam/hydrogen carrier gas. This gas mixture effectively controls the oxygen potential
b) with other fission products
c) with control materials. For a PWR, the silver, indium and cadmium constituents of the control rods are
particularly important, as is the soluble poison boric
acid which is present in the coolant in normal operation

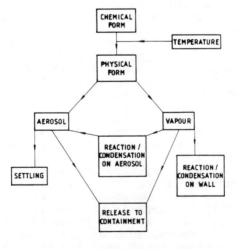

FIG. 3 PROCESSES IMPORTANT IN THE TRANSPORT AND
RETENTION OF RADIONUCLIDES IN THE REACTOR
COOLANT SYSTEM

d) with surfaces, especially the stainless steel surfaces of the reactor vessel and coolant system.

On passing from the very hot regions of the core and upper reactor vessel to cooler zones of the circuit, the vapours of fission products and control materials will become saturated and condense. If they condense on the circuit walls, this represents retention. Alternatively, they may nucleate to form aerosols or condense onto existing aerosols. The behaviour of aerosols is extremely important in determining accident source terms, especially their growth by agglomeration and deposition on surfaces. Heterogeneous reactions between fission product vapours and aerosol surfaces are also possible. Settling out of aerosols can be a significant retention process in the coolant system, and is the dominant retention mechanism in the containment building. Some important chemical interactions of I, Cs, Te and Ru along the transport route are now examined.

Iodine. The evidence for CsI as the chemical form of iodine released from fuel has already been noted. CsI is moderately volatile under accident conditions (B Pt 1280 $^{\circ}$C), and is predicted to be thermodynamically stable in steam/hydrogen atmospheres in the coolant system. At temperatures below about 1000 $^{\circ}$C, condensation of the vapour would become increasingly important, either onto aerosols, or directly onto coolant circuit surfaces. This simple picture of iodine chemistry, which evolved in the aftermath of the TMI accident[12], was considerably perturbed by the discovery that CsI and boric acid react together in the condensed phase over a wide temperature range to release gaseous hydrogen iodide[14]. The reaction is complex, producing a mixture of caesium borates, but can be conventionally represented as:-

$$CsI + HBO_2 \rightarrow HI + CsBO_2 \qquad (2)$$

In certain PWR accidents, boric acid vapours released from the boiling coolant could co-deposit with CsI in the cooler regions of the circuit, promoting reaction (2). However, HI is itself highly reactive. It reacts rapidly at steel sufaces to form iron and nickel iodides, and also with silver, indium and cadmium which could be present in aerosol form or deposited on surfaces. Thus, iodine not retained in the coolant system is likely to be released to the containment predominantly as metal iodides incorporated in aerosol particles, rather than as HI.

Once in the containment, iodine-bearing aerosols
may act as nuclei for steam condensation, and will ultimately be deposited into pools of water on the containment floor. The soluble portion of the iodine will
enter solution as I^- ions. Such ions, when dissolved
in pools of water, will make little contribution to the
source terms if the containment fails. However, the
containment is an oxidising environment, and I^- may be
oxidised to more volatile iodine species such as
elemental iodine and hypoiodous acid (HOI), which are
capable of partitioning out of solution into the gas
phase. Post-accident conditions provide a situation
where a complex series of radiolytic and thermal reactions can generate stable and transient inorganic
iodine species with oxidation states in the range -1 to
+5. Some of these are listed in Table 4. Volatile organic iodides such as methyl iodide may also be formed
by reactions involving organic materials such as paints
and plastics. Sampling of the containment atmosphere
at TMI some three months after the accident showed that
80-90 % of the airborne iodine at that time was organic
in form, although this represented only 0.02 % of the
total core inventory. The great improvements in the
understanding of the severe accident chemistry of
iodine in the containment since the TMI accident represents one of the major achievements in source term
analysis[15]. To give a fair account of this large topic

Table 4   Iodine oxidation states and well-characterised
          species in solution

Oxidation State	Formula	Name
-1	I^-	Iodide ion
	HI	Hydroiodic acid
0	I_2	Elemental iodine
+0.33	I_3	Tri-iodide ion
+1	OI^-	Hypoiodite ion
	HIO	Hypoiodous acid
+5	IO_3^-	Iodate ion
	HIO_3	Iodic acid

is beyond the scope of the present paper, and details
will be found elsewhere[16]. Suffice it to say that not
only does a good predictive capability now exist, but
that the results of this research provide clear
guidance on the measures which need to be taken to minimise iodine release as part of accident management[18].

   Caesium Because the fission yield of I is much
smaller than that of Cs, CsI can account for less than
10% of the total caesium released from fuel. Other
chemical forms of caesium which need to be considered
are caesium hydroxide and caesium telluride (Cs_2Te).
Cs_2Te may be stable in the condensed state in the
cooler regions (<800 $^{\circ}C$) of the reactor system, and
could account for about 30% of the total caesium.
Caesium hydroxide will be the predominant vapour phase
species, and is moderately volatile (B Pt 1190 $^{\circ}C$) at
the temperatures prevailing in coolant system. The
CsOH vapour released from the core will condense in the
cooler regions, either onto aerosols or onto the walls.
If CsOH is deposited on surfaces where boric acid is
also present, the reactions

$$3CsOH + H_3BO_3 -> Cs_3BO_3 + 3H_2O \qquad (3)$$

$$CsOH + HBO_2 -> CsBO_2 + H_2O \qquad (4)$$

may occur. Reactions (3) and (4) may also occur in the
vapour phase, or on aerosol particles. Caesium borates
are much less volatile than CsOH, and formation of
these compounds may significantly affect the retention
of caesium in the coolant system. In addition, it has
been shown that, at temperatures of 700 - 1000 $^{\circ}C$, CsOH
vapour reacts with the oxidic layer which will be
present on the surfaces of the stainless steel components of the coolant system[14]. This provides a further mechanism for retention of caesium.

   Caesium will be released from the coolant system
to the containment in aerosol form in almost all accident situations. Caesium compounds are soluble, and
these compound will largely determine the hygroscopic
properties of the aerosol particles which influence
steam condensation, and hence particle growth and settling rates. In an intact containment, most of the
caesium will eventually find its way into water pools.

   Tellurium Thermodynamic studies show that Te and
Te_2 will be the most important vapour species for tellurium in the steam/hydrogen atmospheres of the reactor

coolant system. However, there is some evidence that
Te which has reacted with Zircaloy fuel cladding may be
released as tin telluride (SnTe) when the cladding becomes oxidised by steam (reaction (1)). In atmospheres
of near-pure hydrogen, hydrogen telluride (H_2Te) will
be a significant additional species. Elemental tellurium vapour reacts rapidly at stainless steel surfaces at temperatures 700 – 1000 $^{\circ}C$ to form tellurides
of Fe, Ni and Cr. This reaction provides a mechanism
for retention of Te in the hotter regions of the
coolant system. Tellurides may also be formed by surface reactions of Te vapour species with aerosols composed of the control rod materials Ag, In and Cd. On
transport to the cooler parts of the circuit (< 700
$^{\circ}C$), condensed Te phases become increasingly important.
Cs_2Te is predicted to be the thermodynamically
preferred species in the condensed phase, but elemental
tellurium will predominate if caesium is absent. Once
again, condensation may take place onto walls, representing retention, or onto aerosols which may be
transported further through the system. Tellurium
released to the containment will be largely incorporated in aerosols. In this oxidising environment,
species such as TeO_2 may be formed by hydrolysis reactions. There is some evidence that gaseous dimethyl
telluride (($CH_3)_2Te$) may be formed radiolytically.
However, tellurium is expected to exist predominantly
as insoluble solid phases, suspended as aerosols or
deposited on surfaces.

   Ruthenium   Very little release of ruthenium from
fuel is expected unless the reactor vessel fails and
core debris is discharged to the containment. Oxidising conditions may then favour the release of ruthenium
to the containment atmosphere, since the oxides RuO_3
and RuO_4 are much more volatile than elemental Ru or
RuO_2. The conditions in a molten core-concrete interaction, described above, will generally be too reducing
to lead to a large release of Ru. The explosive interaction of core debris with water pools in the containment which might occur following vessel failure could
eject finely-fragmented fuel into the containment atmosphere, providing conditions favourable to the
release of ruthenium. This mechanism for enhanced
oxidation release of Ru was recognised in the RSS, and
methods to gain a detailed understanding of its importance are still being developed.

Chemistry in the TMI and Chernobyl Accidents

The above outline of the role of chemistry in
source term analysis of LWR severe accidents is based
on a large body of experimental and modelling data.
Clearly, there are still many uncertainties, and
several research programmes are in progress aimed at
resolving these. An important question is, how does
this picture compare with what actually happened in the
TMI and Chernobyl accidents? The role of the TMI accident in stimulating severe accident chemistry research
over the last decade has been alluded to already. The
observations of fission product chemistry at TMI have
acted as a benchmark against which much of the
laboratory data and modelling have been compared.
Thus, it is fair to say that we now have a consistent
understanding of the role of chemistry in the TMI accident. The Chernobyl accident, being far removed in nature from anything conceived as possible in a reactor
of Western design, requires a considerable extrapolation from the scenarios discussed above. Nonetheless,
a basic understanding of the main chemistry features of
that accident has emerged[4,17].

All barriers with the environment failed immediately at the onset of the Chernobyl accident,
giving free access of air to the debris of UO_2, zirconium alloy and graphite moderator which had constituted the core. The release of activity extended
over a ten-day period from the 26 April, 1986, by the
end of which at least 50 MCi of activity were present
in the environment. Soviet experts have divided this
period into four release stages. Stage 1 (26 April)
was the initial reactivity excursion. Fuel temperatures in parts of the core rose to around 2800 °C in a
few seconds, before subsiding rapidly to below 2000 °C.
Soviet reports state that 5% of the core inventories of
the volatile fission products were released at this
stage, but some Western estimates suggest that 15-20%
is more likely[4,18]. 0.3-0.4% of the non-volatile
radionuclides were released, including some material
ejected mechanically. This release ratio is consistent
with data discussed earlier on release from overheated
fuel in steam atmospheres. Oxidation of the zirconium
alloy pressure tubes and fuel cladding acted as an important heat source in this Stage. During Stages 2 and
3 (late 26 April-2 May and 3-5 May respectively),
graphite fires in the core debris played a major part
in sustaining temperatures and promoting the escape of
radioactivity. A further 15-20% of the volatile fis-

sion products, and 3-4% of the non-volatiles, were
released. The distribution of non-volatile
radionuclides reflected the composition of whole fuel
(Table 1), except for ruthenium which exhibited a
slight preferential release. Temperatures and chemical
conditions may have been quite diverse through the core
debris. The oxidation of UO_2 in air

$$3UO_2 + O_2 \rightarrow U_3O_8 \qquad (5)$$

was probably the main release mechanism during these
Stages. At temperatures 500-800 $^{\circ}$C, this generates
fine U_3O_8 aerosols, which would be readily carried away
in the convective air streams through the core debris.
Above 1000 $^{\circ}$C, oxidative release of ruthenium becomes
rapid, forming the volatile oxides RuO_3 and RuO_4. Once
released to the cool atmosphere, ruthenium oxide
vapours would condense to form aerosols. This process
was probably responsible for the particles of near-pure
ruthenium composition found in aerosol samples collected in Sweden and elsewhere following the accident.
At still higher temperatures, volatilisation of U_3O_8
may have contributed to release. There has also been
speculation that the carburisation of fuel by graphite

$$UO_2 + 4C \rightarrow UC_2 + 2CO \qquad (6)$$

may have contributed to fission product release through
restructuring of the fuel phases[19]. However, in a ventilated air environment, oxidation is likely to have
been the preferential process. In Stage 4 (6 May), the
activity release was abruptly terminated. This coin-
cided with the injection of nitrogen under the core
debris, the effect of which was presumably to extin-
guish the graphite fires and terminate fuel oxidation.

    In summary, the chemistry which determined the
Chernobyl source terms, dominated by the presence of
oxygen and graphite, has very little relevance to
severe accidents in reactors of Western design. The
fact that it has been possible nonetheless to understand the main chemical processes at Chernobyl reflects
the scope and depth of chemistry research in nuclear
safety.

    4   ACCIDENT CONSEQUENCES

The final step in a severe accident risk analysis,
having defined the source terms, is to assess the consequences. The emphasis to date has been on human

health and economic effects, and the environmental impact has been considered only in connection with these.
Several large computer programmes have been developed
which model the atmospheric dispersion and deposition
of radioactivity released in the environment, and the
radiological and economic consequences, including the
effects of countermeasures[2]. However, rather than dis-
cussing modelling, this review will conclude with a
resume of the environmental consequences of the Chernobyl accident, highlighting the chemical aspects.
Chernobyl demonstrated rather graphically that the consequences of the worst type of severe accident occur on
an international scale. For example, the isotopes
Cs134 and Cs137 have been mainly responsible for the
long-term environmental effects. It has been
estimated[4,18] that of the 20-30 % of the core inventory
released, less than half settled within the territory
of the USSR, and most of the remainder settled in the
rest of Europe. Some 1.5 % of the caesium inventory
settled in Sweden alone.

Chernobyl - Consequences and Countermeasures

USSR. The main human consequences in the USSR have
received wide publicity. The official Soviet death
toll from the accident still stands at the 31 site
recovery workers who died of acute radiation sickness
shortly after the event. Some 130,000 people were
evacuated from a 30 km radius exclusion zone around the
plant within a few days, and this exclusion is still in
force. A lifetime epidemiological survey of the
evacuees is in progress. Less well-known statistics
are that 15000 medical workers were drafted into the
Chernobyl area to examine and monitor a population of 1
million, and that 5.4 million people were given iodine
tablets to block the uptake of I131 by the thyroid[20].

Intensive activities to clean up the site and
decontaminate the undamaged Units 1, 2 and 3 at Chernobyl were undertaken in the weeks and months following
the accident. Mechanical procedures such as collecting
up and disposing of loose fuel fragments and removing
topsoil were used for clearing the worst of the contamination. Few details have been given of the chemical methods employed, and Soviet accounts refer to surface decontamination by 'special solutions', and by the
use of water and steam jets. Some surfaces were decontaminated by application of rapid-setting polymer coatings. Externally, these seem to have been widely used
to stabilise surfaces and suppress the resuspension of

activity. Internally, they may have been subsequently
peeled away to lift off activity. The heavily-damaged
Unit 4 was entombed in a massive concrete
'sarcophagus', which was completed by December 1986.
This provided both radiation shielding, and controlled
and filtered ventilation of the the core debris to
remove decay heat. By October of 1986, decontamination
had progressed to the point that Unit 1 could be re-
started. Units 2 and 3 resumed operation in December
1986 and December 1987, respectively.

Beginning during the period of discharge from Unit
4, and continuing up to the present time, there has
been extensive monitoring of the airborne and deposited
radioactivity by Soviet scientists. This has been most
intensive within and just outside the 30 km zone.
Isotopes of Cs, Sr, Pu, Zr, Nb, Ru and La were identified as contributors to the persistent radiation
levels, with Cs134 and 137 generally dominant. New
radiochemical methods were developed for the analysis
of soil, water and aerosol samples, particularly necessary for Sr and Pu measurements. A method of remote
monitoring of radiation levels by ultraviolet lumines-
cence of air was also developed. Average surface activity levels within the 30 km zone fell by a factor of
55 during the year following the accident. Radioactive
decay contributed most of this fall, but transport into
the soil was an additional factor, and migration depths
up to 1.2 m were recorded. Significant surface
transport of activity by wind-borne dust and organic
matter was also observed in the early aftermath of the
accident.

Some 5 million acres of land outside the 30 km
zone have been classed as contaminated. The contamination levels were heavily influenced by local rainfall
during the period of discharge, which washed out airborne activity and created some hotspots. Between 50 %
and 70 % of the contaminated territory is occupied by
forests, and there are also large areas of marshland.
On average, only about 25 % of the land is in agricultural use. Three levels of response have been adopted
with respect to agriculture:

- complete ban on all food production on land with
Cs137 contamination at levels 40 Ci/km^2 and higher.

- changed agricultural techniques for land contaminated
at intermediate levels. For arable land, this has included the addition of radionuclide adsorbers (clay

suspensions and zeolites) to the surface layers, followed by deep ploughing. Binding of Cs by these
materials is particularly effective. On some pasture
land, a technique of heavy application of mineral fer-
tilisers and liming has been adopted. Potassium and
calcium in these materials have acted as competitors
for uptake with the radioisotopes of caesium and strontium from the fall-out, thus reducing crop contamination.

- normal consumption of food from land contaminated
below the level prescribed in safety standards, with
spot monitoring.

Only limited reports of ecological damage have emerged
from the USSR. Extensive soil sterilisation was observed within the 30 km zone, but recent reports suggest that micro-organic activity has returned to pre-
accident levels in much of the zone. Populations of
small mammals and birds are also reported to have
recovered. Pines proved to be the most susceptible of
tree species to radiation damage and mutation. 1000
acres of pine forest are reported to have been
destroyed, but this is probably an underestimate.
Deciduous trees have been much less affected, although
there have been reports of mutations producing over-
sized leaves in species such as oak.
A major concern of the Soviet authorities immediately
after the accident was the contamination of water supplies. The Pripyat River passes within 2 km of Chernobyl, and ultimately feeds the Kiev reservoir, which
is the chief water supply of that city. Extensive
water monitoring programmes were established covering
all rivers and streams in the contaminated zone. These
revealed much lower levels of contamination than an-
ticipated, even in the Pripyat River two months after
the accident. Strong retention of radionuclides by
sorption onto soil presumably prevented transport of
deposited activity by surface and ground water run-off.
Approximately 70 % of the caesium contamination in the
waterways was found to be adsorbed onto sedimentary
material. During the winter of 1986/87, a system of
100 protective and filtering dams was constructed on
the waterways feeding the Kiev reservoir to cope with
the influx from the spring floods. In the event, no
rise in contamination of the reservoir water above the
levels of the previous autumn was found. Overall, Cs137
contamination fell by a factor of about 20 between July
1986 and May 1987, to a level of around 1 Bq/litre.
The levels remained well within the prescribed limits

acceptable for drinking throughout the post-accident
period.

UK   The consequences of the Chernobyl accident in the
UK were very low. The average lifetime incremental
radiation dose to the individual (0.05 mSv) has been
equated with the additional dose acquired from taking a
three-week holiday in an area of high natural background radiation, such as Cornwall[4]. Highest ground
contamination occurred in the areas of heavy rainfall
as the radioactive cloud passed over Britain on the 2-3
May, 1986. These include parts of North Wales, Cumbria
and Scotland. Wide publicity has been given to the
restrictions on the movement and slaughter of sheep
from these areas, imposed shortly afterwards and still
in force in some localities. It should be emphasised
that this is a precautionary measure, and that the
level of Cs137 contamination in sheepmeat at which action was triggered (1000 Bq/kg) is well below that normally accepted as safe. The ready transfer of caesium
to sheep was due to the poor binding of the element to
the humic, acidic soil in upland pasture, resulting in
strong uptake by grazing vegetation. An extensive survey of the effects of radiocaesium deposition in the
UK[21] has shown an accelerating decline in average surface contamination over the year following the accident, by factors of up to 50. Significant transfer to
game and wildlife through foodchains has been measured,
but in no case have levels been reached in edible
species which would pose a risk to even a voracious
consumer.

        5 SUMMARY

The severe consequences of the Chernobyl accident have
had such an impact on the public conciousness as to
present some difficulties in retaining a perspective on
the risks associated with severe reactor accidents.
Such a perspective is necessary, and it is worth
reiterating that circumstances of the Chernobyl accident were unique to the Soviet Union at that time.
Chernobyl has brought about a radical revision of the
Soviet approach to nuclear safety, including greatly
increased contacts with the West, and a desire to adopt
Western safety practices and standards. The extensive
safety measures taken in the design, construction and
operation of Western nuclear power plant are such as to
render the risk of death or injury to the individual
from a severe accident several orders of magnitude
smaller than other accident risks of everyday life.

Moreover, even in the West, the Chernobyl accident has acted as a stimulus to the nuclear industry to exhaustively re-examine its safety principles and practices, for any further improvements which might be identified.

The TMI accident was severe in terms of plant damage, but the health and environmental consequences were negligible. From the chemist's viewpoint, one of the more important outcomes of the TMI accident was to promote an increased recognition of the importance of chemistry in the analysis of severe accident source terms. A high level of research activity on this topic has been maintained through the 1980s, which has greatly increased the detailed understanding of how radionuclides will behave under severe accident conditions. Many new, and occasionally unsuspected, chemical interactions have been revealed and investigated. Some features of the chemistry of the fission products iodine, caesium, tellurium and ruthenium have been described here. It should be emphasised that much of this work has been valuable as a contribution to the body of fundamental chemical knowledge, and its applicability is not confined to accident analysis. This is particularly the case in the areas of high temperature chemistry, chemical thermodynamics, and radiolytic chemistry. As a result of this work, we are now in a position to make much sounder predictions of radioactivity releases in postulated severe accidents. However, there are still some important uncertainties, and research to resolve these must continue.

The conditions in the Chernobyl accident were far removed from any which have been predicted for a severe accident in a Western design of reactor. Nevertheless, the breadth of the chemical understanding underlying Western safety studies is such that main features of the Chernobyl source terms can be adequately explained. Studies of the consequences of the Chernobyl accident have been intensive since it occurred, and will continue for many years to come. These include extensive investigations of the environmental impact, and optimisation of clean-up and recovery measures. Earlier this year, an International Research Institute was established near Chernobyl to facilitate this work. Thus, much is being learned from the world's worst nuclear power disaster on how to minimise the health and environmental consequences of severe accidents. However, this is ancillary to the nuclear industry's real concern, which is to ensure that such an accident will never happen again.

ACKNOWLEDGEMENT

This work was supported by the UKAEA General Nuclear
Safety Research Programme funded by the Department of
Energy.

REFERENCES

1.   USNRC. Reactor Safety Study. An Assessment of
     the Accident Risks in US Commercial Nuclear Power
     Plants. WASH-1400. 1975.

2.   USNRC. Severe Accident Risks: An Assessment for
     Five US Nuclear Power Plants. NUREG-1150
     Volumes 1-3. 1989.

3.   Sizewell B Probabilistic Safety Study. WCAP-9991
     Westinghouse Electric Corporation. 1982.

4.   J.H. Gittus et al. 'The Chernobyl Accident and
     its Consequences'. 2nd Edition. UKAEA Report
     NOR 4200. 1988.

5.   OECD. 'Chernobyl and the Safety of Nuclear Reac-
     tors in OECD Countries'. Report by an NEA Group
     of Experts. 1987.

6.   USSR State Committee on the Utilisation of Atomic
     Energy. 'The Accident at Chernobyl Nuclear Power
     Plant and its Consequences'. IAEA Experts Meet-
     ing, Vienna, 25-29 Augusut 1986.

7.   P.N. Clough. 'Source Terms and the Chernobyl
     Accident'. Nuclear Safety after Three Mile Island
     and Chernobyl (Ed. G M Ballard). Elsevier Applied
     Science. 1988 p306.

8.   J.H. Gittus et al. PWR Degraded Core Analysis
     Chapter 7. UKAEA Report ND-R-610(S). 1982.

9.   M. Rogovin and G.J. Frampton. 'Three Mile Island.
     A Report to the Commissioners and the Public'. US
     Nuclear Regulatory Commission Special Inquiry
     Group, 1980.

10. D.J. Wren. 'Kinetics of Iodine and Caesium Reac-
     tions in the CANDU Reactor Primary Heat Transport
     System Under Reactor Accident Conditions'. Atomic
     Energy of Canada Ltd Report AECL-7781. 1983.

11. D.O. Campbell, A.P. Malinauskas and W.R. Stratton, Nuclear Technology, 1981, 53, 11.

12. USNRC. 'The Technical Bases for Estimating Fission Product Behaviour during LWR Accidents'. NUREG-0772, 1981.

13. US National Research Council. Workshop on Chemical Processes and Products in Severe Nuclear Reactor Accidents. December 9-12, 1987. Captiva Island, Florida.

14. B.R. Bowsher. 'Fission Product Chemistry and Aerosol Behaviour in the Primary Circuit of a PWR Under Severe Accident Conditions'. UKAEA Report AEEW-R1982. 1987.

15. OECD. 'Proceedings of the Second CSNI Workshop on Iodine Chemistry in Reactor Safety'. Atomic Energy of Canada Ltd Report AECL-9923. 1989.

16. D.J. Wren et al. 'A Review of Iodine Chemistry Under Severe Accident Conditions'. Atomic Energy of Canada Ltd Report AECL-9089. 1987.

17. P.N. Clough. 'The Chernobyl Accident - Source Terms and Related Characteristics'. Chernobyl - A Technical Appraisal. British Nuclear Energy Society, London, 1987.

18. L.R. Anspaugh, R.J. Catlin and M Goldman. Science, 1988, 242, 1513.

19. D.A. Powers, T.S. Kress and M.W. Jankowski. Nuclear Safety, 1987, 28, 10.

20. V.G. Asmolov et al. 'The Chernobyl Nuclear Power Station Accident: One Year Afterwards'. IAEA Conference on Nuclear Power Performance and Safety, 28 September - 2 October 1987 Vienna. IAEA-CN-48/63.

21. A.D. Horrill, V.P.W. Lowe and G. Howson. 'Chernobyl Fallout in Great Britain'. Department of the Environment Report DOE/RW/88.101. 1988.

The Potential Radiological Consequences of Deferring the Final Dismantling of a Magnox Nuclear Power Station

P. B. Woollam

NUCLEAR ELECTRIC PLC, TECHNOLOGY DIVISION, BERKELEY NUCLEAR
LABORATORIES, GLOUCESTERSHIRE GL13 9PB, UK

1. INTRODUCTION

When it is no longer economic to maintain Nuclear
Electric's power stations to their current high safety
standards, the reactors will be safely shutdown and
decommissioned. Nuclear Electric is developing detailed
engineering plans for the decommissioning of its
earliest nuclear plant: the steel pressure vessel Magnox
reactors. This work will follow the three
internationally accepted stages: the first two will
involve defuelling the reactors and then dismantling all
plant and buildings external to the 2 m thick reinforced
concrete bioshields. Almost all of the original site
would then be available for re-use if required. The
third stage will be the dismantling of the two reactor
islands, including the concrete shields, steel pressure
vessels and graphite cores.

Stages 1 and 2 of decommissioning could be complete
within 10 to 15 y, including a 5 y period to defuel the
reactors: this will remove 99.99% of the radioactivity
which was on site whilst the reactors were running. The
timing of Stage 3, complete reactor island dismantling,
will depend on a number of factors, including balancing
the need to re-use the site against the advantages of
allowing residual radioactivity in the structure to
decay.

If final dismantling is deferred, it is intended
that the residual reactor buildings would be maintained

in a sealed, weathertight condition, and it is confidently expected that this can be achieved. This paper reports Nuclear Electric investigations of the potential consequences of any failure in this containment during the deferral period using, as a demonstration of the robustness of the radiological case for deferment, some extreme hypothetical scenarios in which the nuclear islands are assumed to have collapsed. No initiating event can be foreseen for such an occurrence and, clearly, this paper is not intended to imply that such a situation would ever be allowed to develop. On purely radiological grounds it would be most sensible to defer final dismantling for a period of up to about 130 y to take advantage of natural radioactive decay. However, again to demonstrate the robustness of the case for deferment, this investigation considers radiological consequences over a time period which is roughly one order of magnitude both greater and smaller than the 130 y optimum.

If final dismantling of Nuclear Electric's Magnox reactors were to be deferred, the dose commitment to dismantling staff would be reduced as far as practicable, thus complying with the spirit of the ALARP principle. Dose rates inside a Magnox reactor, as a function of time, are shown in Figure 1. It would also reduce the need to rely on robotic systems for dismantling and the cost savings to the electricity consumers would be very significant, particularly totalling the savings over all Nuclear Electric's Magnox stations. A further possible reason for deferring Stage 3 decommissioning might be that, for political rather than technical or safety reasons, no disposal routes were immediately available for the radioactive waste. In this eventuality the material would remain on site, within the existing secure containment afforded by the reinforced bioshields.

During any deferral period Nuclear Electric would remain responsible for the site and for the residual radioactive material contained within the existing shields. These have walls which are 2 m thick and 6 m thick roofs. It is easily shown that the dose equivalent rate to a person standing just outside the shield concrete, resulting from residual radioactivity in the structure is about 10^{-5} of that from the local natural background. Because of the construction of the concrete shields, which are massively reinforced on both faces, it is not feasible for unauthorised access to be gained to the reactor structures without the use of very sophisticated cutting technology. For the same reason the structures would not present a worthwhile target for attack by extremist groups.

However, if a decommissioned Magnox reactor were to be left for an extended period before dismantling, the possibility that residual radioactivity from the structure might be transferred to the groundwater system must be considered. The most likely route for radioactive materials to transport into the groundwater is rain leakage through discontinuities which might eventually form in the top shield, followed by corrosion of the radioactive structure and subsequent leakage of contaminated water through similar discontinuities in the foundation raft. The principal objective in assessing such safety issues is to examine the engineering standards required of the structure during the period for which final dismantling is deferred. The degree of containment which is necessary, and the time period over which such containment might be required, are primary inputs to the process of deciding whether it is radiologically acceptable to defer Stage 3 dismantling. Dungeness in Kent presents a limiting case in such an assessment since it is the only Nuclear Electric site built on an aquifer from which large quantities of drinking water are extracted for the nearby population.

Nuclear Electric's investigation set out to determine the radiation exposure which might result from the transfer of radioactive materials from the two Magnox reactor structures to the Dungeness aquifer in the event that final dismantling were to be deferred and the reactor containment to fail. It also assesses the radiological consequences of gaseous discharges from oxidation of the graphite cores for those radioactive isotopes where atmospheric release is likely to be the major route by which radioactivity returns to man.

2. BASIS OF THE ASSESSMENT

The radioactivity inventory of the Dungeness 'A' Magnox reactors is based on a series of complex neutron transport calculations (1) confirmed by measurement (2). A careful and systematic check of all 2600 known isotopes (3) showed that nothing significant had been overlooked in determining the inventory. The long term

peak dose rates inside the structure 130 y after final
reactor shutdown are expected to be 3 μSv h^{-1}, allowing
man-access in excess of 30 h week^{-1} to set up and
maintain dismantling equipment. These peak dose rates
occur in the interspace between the pressure vessel and
the core restraint structure and come from the isotopes
Ag-108m and Nb-94 (see Figure 1). The total radioactive
inventory of a reactor 100 y after shutdown is
dominated, in simple numerical terms, by C-14 and
Ni-63. Both isotopes are beta-emitters with low energy
endpoints; their contribution to the whole body dose
rate experienced by dismantling workers is negligible.
The only plant likely to be contaminated with actinides
or fission products is the spent fuel storage pond and
the radioactive
effluent treatment
plant. These
facilities are
expected to be
decontaminated and
dismantled during
Stage 2 of
decommissioning, as
soon as practicable
after all the fuel has
been dispatched for
reprocessing.
Rainwater ingress to
the reactor island is
modelled by assuming
that the water leakage
rate increases
linearly with time
after the structure is
first breached. The
model has been applied
to extreme situations
and allows for
different breaching
times for both the
concrete shields and
the steel pressure
vessel, together with
different water ingress
rates and, eventually,
different times at
which both parts of the
structure finally
collapse. It also

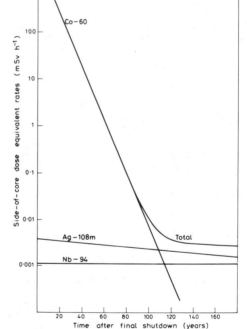

Figure 1: Reduction with time
in the maximum dose
rate to a dismantling
engineer.

assumes that both reactor islands breach and collapse simultaneously. This extreme treatment of the time dependence of possible structural failure allows an assessment to be made of the engineering standards necessary to contain the radio-logical consequences of rainwater ingress below particular limits.

The release to water of many of the radionuclides present in the nuclear island structures is limited by their solubility in **Figure 1:** Reduction with time in water (which is strongly dependent on the local aqueous chemistry), and not by the bulk corrosion rate of the materials in which they were activated. For these particular radionuclides this assessment has therefore applied the concept of a solubility limited release rate. The release rates to water of three other isotopes (C-14 and H-3, for which gaseous transfer to man is dominant and Cl-36, which is highly soluble) are however assumed to depend on their leach rates (4): these are not solubility limited. The dominant source of these isotopes is the graphite moderator.

Groundwater transport parameters used in the model to determine radioactivity transport within the aquifer are available from detailed hydrogeological surveys, undertaken by the Southern Water Authority (5), of the

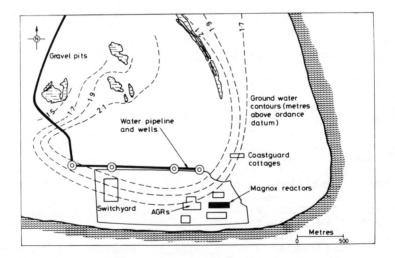

Figure 2: The Dungeness area, showing the reactors relative to the wells and the houses.

aquifer on which Dungeness Power Station is sited. The nearest well is about 500 m from the Magnox reactors and the whole 14 km aquifer system in 1982 supplied 2.3×10^6 m^3 y^{-1} to properties along the south coast of England. Figure 2 shows the location of the wells in relation to the power station. An important parameter in groundwater radioactivity transport assessments is the retardation coefficient which describes the ability of the rock medium to hold up a particular radionuclide until it has decayed. This parameter is critically dependent on the chemistry of the rock and the water, thus measurements have been made by Nuclear Electric to determine retardation coefficients for several important isotopes under the conditions prevailing in the aquifer.

Not all of the radioactive isotopes in the decommissioned structures have the aquifer as their principal route for returning to man. C-14 and H-3, predominantly found in the graphite, will be released to the atmosphere rather then the aquifer. A detailed methodology is available (6) for assessing the annual dose to the most exposed individual from continuous aerial releases of these two isotopes. Site specific studies show that the most exposed individuals live in ten coastguard cottages to the NNE of the Dungeness site.

Further details of the overall methodologies used in this work are presented elsewhere (7).

3. RADIOLOGICAL CONSEQUENCES

Using the data outlined above, the radiological consequences of deferring Stage 3 decommissioning of Dungeness 'A' nuclear power station were determined for three basic scenarios:

i)   That 15 y after final shutdown the nuclear island concrete starts to leak at a slow rate which increases with time until the shields and vessels eventually collapse at some distant, but defined, point in the future.

ii)   That following Stage 2 dismantling the nuclear islands will receive a treatment, or cover, of such quality that there will be no possibility of leaks over an initial extended period, taken here to be 500 y. After this time it is assumed that the degradation processes used in scenario 1 will apply, leading to eventual collapse. Comparison between the consequences of scenarios 1 and 2 will show whether it is radiologically justifiable to spend significant sums on superlative engineered protection at the start of the deferment period.

iii) That the entire nuclear island, pressure vessel and core collapse completely at the end of the Stage 2 process, 15 y after final shutdown: this clearly is the maximum possible fault situation. It has been considered here even though no credible initiating event can be foreseen.

The radiological consequences of these scenarios were assessed for a period of 1500 y following the time when the bioshield concrete is first breached. It is intended that the results should be best estimates, thus all the important input data specifically apply to the Dungeness case.

3.1 Scenario 1: Early Leakage

Figure 3 shows the total individual dose rate to people living in the coastguard cottages and drinking the aquifer water if the bioshields and pressure vessels of

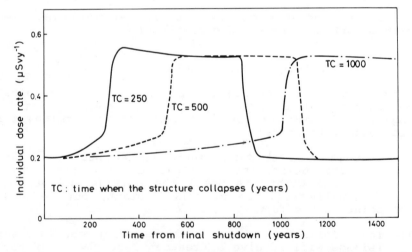

Figure 3: Variation in individual dose rate with time if
rainwater starts to leak into the reactors
15 y after shutdown.

the two reactors are breached 15 y after shutdown and eventually collapse at times between 250 and 2000 y later. A peak dose rate of 0.5 μSv y^{-1} exists for a period of 500 y following the time when the system collapses. Outside this peak the dose rates are 0.2 μSv y^{-1}.

For the case of collapse after 500 y, Figure 4 shows how the dose rate depends on the various isotopes leaching from the reactors. The long term dose rate of 0.2 μSv y^{-1} comes principally from Cl-36 released to the aquifer and C-14 released to the air: both isotopes originate in the graphite. The peak in the dose rate curve is formed by Ca-41, again leached from the graphite at a rate limited by its solubility. Contributions from the long lived metallic species Ni-59, Ni-63, Nb-94 and Ag-108m are much lower, limited by their solubilities and, for Ag-108m and Ni-63, their half lives.

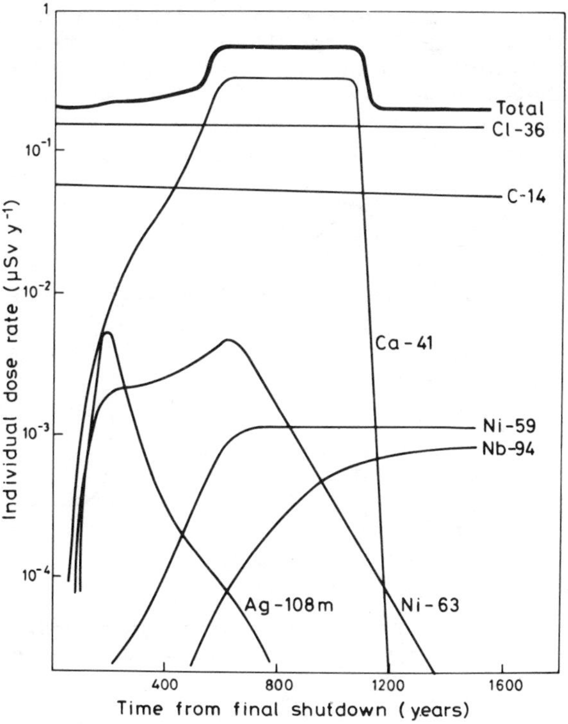

3.2   Scenario 2: Late Leakage

This scenario is identical to scenario 1, except that it is assumed that some exceptional engineering treatment has been applied to the nuclear islands to prevent water

Figure 4:   Contributions to dose rate from different isotopes.

ingress until 500 y after shutdown. The results show
that the dose rate picture is very similar to the first
scenario. This of course is not surprising, since from
the previous discussion it is clear that the dose rates
are controlled by the long lived isotopes C-14, Cl-36
and Ca-41. Over any period for which total and complete
containment might be credible the radioactivities of
these isotopes will scarcely change.

3.3   Scenario 3:   Early Complete Collapse

In the third scenario, the nuclear islands assumed to
collapse completely following Stage 2 of
decommissioning, thus allowing rainwater access to the
entire structure 15 y after shutdown. It is not
possible to foresee a credible initiating event for this
scenario. The resultant dose rates are shown in Figure
5, where it can be seen that the peak occurs 130 y after
shutdown at 0.7 μSv y^{-1}. The dose rate then drops

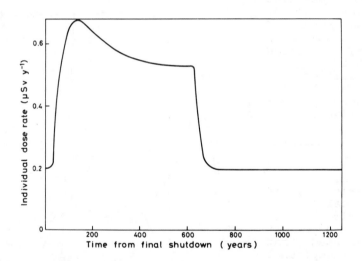

__Figure 5:__ Variation in individual dose rate with time if
the whole reactor collapses completely 15 y
after shutdown.

slowly to the level found in the previous scenarios, 0.5 μSv y^{-1}, until 700 y after shutdown it falls to the long term level of 0.2 μSv y^{-1}. The peaked form of the dose rate variation comes from the contributions, at 135 y decay, of Ni-63 (0.2 μSv y^{-1}) and Ag-108m (0.02 μSv y^{-1}) which have decayed before the major releases start in the previous two scenarios. It is clear that, even with the bioshields collapsing just 15 y after shutdown,the short lived nuclides such as Co-60, Fe-55 and H-3 do not contribute any major increase in dose rate to the local population.

4. ENGINEERING CONSIDERATIONS AND UNCERTAINTIES

The best estimate data used here lead to the very clear conclusion that the radiological consequences of deferring Stage 3 dismantling are very slight.

Considering first the release of radioactivity to the air (which is independent of the various water ingress and structure collapse scenarios), we find resultant dose rates of 0.1 μSv y^{-1}, from the two Magnox reactors, to the most exposed individual who lives in the coastguard cottages to the NNE of the power station site. This is almost entirely due to C-14 from the graphite. Release rates of C-14 to the air are based on measurements of the transfer of this isotope from Dungeness moderator graphite samples to water, under a variety of chemical conditions. There is very little uncertainty on the specific activity of C-14 in moderator graphite. However, the natural C-14 dose rate to the people in the cottages is about 50 times higher than the dose rate from airborne releases from the two decommissioned reactors.

If the two Magnox reactors both collapsed completely just 15 y after shutdown, the best estimate peak dose rate to water consumers would be 0.6 μSv y^{-1}, 120 y later. A similar collapse after 250 y decay would lead to a peak dose rate of 0.5 μSv y^{-1}. The peak dose rate from the decommissioned reactors, through the aquifer route, is controlled by Cl-36 and Ca-41, both originating in the graphite cores. Ca-41 releases are solubility limited and so it is unlikely that this isotope could be released faster than the model predicts. It is assumed that Cl-36 is released at the same rate as the bulk corrosion rate of the graphite (4). The bulk corrosion rate was determined, as

described earlier, by measuring the leach rate to water
of C-14 from samples of irradiated Dungeness graphite
moderator.   At this rate Cl-36 contributes 0.14 μSv y^{-1}
to the dose rate to water consumers, or some 25% of the
total dose rate from the two decommissioned reactors.

From Figure 4 it can be seen that uncertainties in
the parameters relating to any of the other isotopes
considered here would need to be very large to affect
significantly the outcome of this study.   Indeed most of
the isotopes considered lead to dose rates far below the
lowest (10^{-5} μSv y^{-1}) shown in Figure 4.

All the hydrogeological parameters used in this
work have been measured for the aquifer: the results
presented here are best estimates based on real data.
The results are relatively independent of some of the
parameters,   for   example   the   ground   water   velocity,
because the half lives of the critical isotopes are so
long.   However, it is implicitly assumed that the water
table slopes from reactor to well: in fact in recent
years this has been the case only for very limited
periods,   certainly   not   continuously.   Figure   2   shows
typical groundwater levels around the Dungeness site.
Clearly when the slope is seawards no radioactivity will
reach the well but, over the time periods considered
here, we cannot be sure that the water table will not
permanently slope, as assumed here, in the critical
direction from reactor to well.

Comparison between the various scenarios described
in this report shows that there is little to be gained
radiologically   by   cladding   the   residual   nuclear
buildings   or   by   extensive   engineered   sealing
techniques.   The peak dose rates resulting from collapse
after a 500 y totally sealed period are the same as
those following a collapse after the reactor islands
have leaked continuously following Stage 2 dismantling.
Indeed the dose rates following a total collapse just
15 y after shutdown are only a little higher than those
found   following   a   collapse   many   centuries   into   the
future.   This leads to the fundamental conclusion that
the   very   long   lived   isotopes   considered   here   will
eventually transport back to man from any engineered
surface repository or store.   In the deferred Stage 3
state, the majority of the radioactivity is contained
within a 100 mm thick steel shell inside a 2000 mm thick
reinforced concrete structure.   This is an engineering
system   which   should   be   comparable,   in   the   very   long

term, with near surface waste disposal sites. The radiological consequences of waste management following decommissioning will not therefore be reduced in practical terms by dismantling the reactor structures, putting them into boxes and placing those boxes in a purpose built near surface repository. Thus the only requirement on the residual structures is that any degradation during a 130 y period following shutdown does not result in holes sufficiently large to allow people to enter the buildings.

The boiler units from the Magnox reactors are likely to be stored on site during the period for which Stage 3 dismantling is deferred. Although the radiological consequences of this option have not been explicitly considered here, the resultant dose rates from leaching and radioactivity transfer to the aquifer can be compared with those from the residual nuclear buildings. The isotopic distribution of material in the boilers is broadly similar to that within the structure, since this is its origin, having been activated within the core and transferred to the boilers with the coolant gas. It is expected that the boiler radioactivity levels will be about 0.01% of those in the nuclear island, thus the contribution to the total dose rate due to water and airborne radioactivity transport to the local population will also be about 0.01%. The boilers are contained in 75 mm thick steel shells, in which they will be stored, so their degradation rate will be very similar to that of the 100 mm thick reactor pressure vessel.

Early intrusion into the concrete shields will clearly result in unacceptable dose rates; however very sophisticated cutting techniques would be required to gain entry and, because the residual reactor structures are above ground, it cannot be argued that unintentional intrusion is likely, at least for the periods for which institutional controls are effective. Unauthorised entry would lead to a situation where, 130 y after shutdown, an individual would need to spend 500 hours each year in the vicinity of the most active components to receive an exposure equal to that from natural background. After about 500 y only Nb-94 will be important and this time will rise to 1300 hours per year.

5.   SUMMARY

The potential radiological consequences of deferring the
final dismantling of the two reactors of the Dungeness
'A' Magnox power station are extremely small.   If the
concrete bioshield structures and steel pressure vessels
both start to leak 15 y after shutdown, at a slow but
increasing   rate   which   leads   to   the   eventual   entire
collapse of both nuclear islands, the peak individual
ingestion   and   inhalation   dose   rates   to   the   local
population   would   be   0.5 μSv   y^{-1}.   Even   if   the   two
complete nuclear islands collapsed completely just 15 y
after final reactor shutdown allowing rainwater access
to the entire structure, peak dose rates to the local
population   would   be   only   0.7 μSv   y^{-1}.   The peak dose
rate   results   primarily   from   radioactivity   returned   to
man through the aquifer on which Dungeness power station
is   built.   There   is   no   radiological   advantage   in
applying extensive engineered sealing technology to the
reactors at the start of the deferment period.   Even if
such technology prevented water ingress completely for
500 y, the eventual radiological consequences when the
system did leak would be the same as those from early
leakage because the half lives of the isotopes involved
are very long.

The gamma dose rate at the outside of the concrete
shields,   resulting   from   residual   radioactivity   inside
the nuclear island, is estimated to be 0.02 μSv y^{-1},
15 y after shutdown.   There is therefore no radiological
reason to control public access to the outside of the
two residual nuclear islands, and the only engineering
constraint   on   the   residual   structures   is   that   they
should not degrade in such a manner that public access
inside the buildings becomes possible during the first
130 y following shutdown.

ACKNOWLEDGEMENT

This paper is published with the permission of Nuclear
Electric plc.

REFERENCES

(1)   P   B   Woollam   and   I   G   Pugh,   'Neutron   Induced
        Activation, Waste Disposal and Radiation Levels for
        the   Reactor   Island   Structure   of   a   Decommissioned
        Magnox Power Station', IAEA-SM-234/10, Vienna, 1979,

(2)   P B Woollam, CEGB Report RD/B/N4231, 1978,

(3)   L D Felstead and P B Woollam, CEGB Report
      TPRD/B/0386/N84

(4)   I F White et al, 'Assessment of Management Modes
      for Graphite from Reactor Decommissioning' CEC
      report EUR 9232, 1984,

(5)   Southern Water Authority, 'Joint Report of the
      Denge Hydrogeological Study', 1984

(6)   S Nair, CEGB Report RD/B/N4668, 1979

(7)   P B Woollam, CEGB Report TPRD/B/0871/R87, 1987

Fusion Reactors and the Environment

R. Hancox
AEA FUSION, CULHAM LABORATORY, ABINGDON, OXFORDSHIRE OX14 3DB,
UK

1 INTRODUCTION

Fusion power, based on the nuclear fusion of light
elements to yield a net gain of energy, is being studied
in many countries as a potential addition to fission power
as a long term energy source. The attraction of fusion
power is that it uses fuels, such as deuterium and lithium
which are abundant, and that neither the fuels nor most of
the reaction products are radioactive. Thus fusion power
has the potential to extend the world's energy resources
in a way which is environmentally attractive.

At the present time, the main aim of fusion research
is to achieve the necessary conditions in high temperature
plasmas to allow nuclear reactions to take place. This
involves the magnetic or inertial confinement of hydrogen
isotopes at temperatures of around 100 M°C for times long
enough for a net energy gain. Progress in this research
has been sufficiently encouraging in recent years that
extensive design studies are being undertaken for experimental fusion reactors which could be built by the turn of
the century. Reactor studies are also addressing ques-
tions of the feasibility of power reactors, together with
preliminary assessments of their economics, safety and
environmental advantages.

Several studies of the environmental impact of fusion
power have already been undertaken. One of the first of
these, published in 1975, was the report of a study group
at the Culham Laboratory,[1] which identified the main
potential hazards as being tritium and activated constructional materials. The tritium is bred in a nuclear blan-

ket which surrounds the reacting plasma of a D-T fuelled reactor and is recirculated as the fuel, and the constructional materials become activated as a result of neutrons generated by the fusion reactions. Whilst concluding that fusion reactors would have several features which offered potential environmental advantages, the report considered that more detailed reactor designs were necessary before these could be quantified. In the same year the IAEA published a review[2] which came to similar conclusions. In the following year a study by the International Institute for Applied Systems Analysis[3] compared the characteristics of fusion and fast breeder reactors.

In 1985, a second study at Culham Laboratory[4] came to similar conclusions stating that, on the basis of several conceptual reactor designs, fusion reactors appeared to have an inherently lower hazard potential than fission reactors although the realization of this potential had yet to be demonstrated. Amongst the intrinsic advantages of fusion were listed the absence of actinides and volatile fission products, the relatively short half lives of the tritium and activated material, and the absence of a separately sited reprocessing plant.

In 1986 an American committee undertook a detailed study of the environmental, safety and economic aspects of magnetic fusion energy (ESECOM). The final report was published[5] in 1989. As a basis for this study, outline designs were developed for ten tokamak and Reversed Field Pinch reactors, including D-T and D-^{3}He fuelled reactors and hybrid fission-fusion reactors. These cases were developed and analysed to permit exploration of a wide range of materials and power densities, and to provide several quantitative comparisons with fission reactors.

In 1986, the Commission of the European Communities published a review of the environmental impact and economic prospects of nuclear fusion.[6] It considered the hazards from tritium and structural radioactivity, both in relation to normal operation and accidents. This report was discussed at a public meeting in November 1987, organized by the Committee on Energy Research and Technology of the European Parliament, and was the first activity undertaken by the Parliament's Scientific and Technological Options Assessment Panel.

In its subsequent approval of the Community's pluri-annual fusion programme, the European Parliament requested that such a review be undertaken again, and this process has now started. As a first step, the Commission established a Study Group to prepare a report on the environmental, safety and economic potential of fusion power. This EEF Study Group commissioned a series of papers which they used as the basis of their report. Several of these reviews are now available, together with the report of the Study Group[7] which was published in December 1989. Some of the information in this paper is based on these reports, which are the latest available consideration of the environmental impact of fusion.

Within the European fusion programme, only the magnetic confinement of plasmas is considered. Furthermore, the majority of work is based on a single confinement geometry, the tokamak, and the majority of reactor studies have been based on this configuration. Since most of the environmental implications of fusion are independent of the means of plasma confinement or of the particular geometry, only tokamak systems are considered in this review. In the American fusion programme, both magnetic and inertial confinement are pursued, but the conclusions of safety and environmental studies are essentially the same. The only exception is the hybrid reactor, in which the plasma is surrounded by a blanket containing fissile material with the objective of breeding fuel which may be used in fission reactors. Such hybrid reactors have some economic advantages, and the safety advantage relative to fission reactors that the blanket is sub-critical, but otherwise do not appear to have substantial advantages over a fast fission breeder reactor.

The majority of reactor studies are also limited to consideration of the deuterium-tritium (D-T) fuel cycle. This is the easiest fusion reaction to achieve, but has the two major disadvantages mentioned above of tritium and induced radioactivity in the constructional materials. It will certainly be the basis of the first generation of fusion reactors. There are alternative fuel cycles, of which the most attractive is $D-^3He$, for which tritium is not required and no neutrons are produced to activate the structure. However, this reaction is far more difficult to achieve, and is not considered in this review.

In the following sections an introduction to fusion reactors is given, and then various aspects of their environmental impact are considered. In this context, the implications inside the reactor site of maintenance operations or accidents are not considered.

2   FUSION REACTORS

A D-T fusion reactor will operate with deuterium and
tritium fuel, confined at temperatures of 100 M°C.   The
reactions produce α particles of energy 3.5 MeV and
neutrons of energy 14.1 MeV.   The α particles remain
trapped in the plasma and are the primary source of
heating for the fuel.   The plasma is transparent to the
neutrons, which are absorbed in a surrounding blanket.
This blanket will contain lithium so that ^{6}Li(n,α)T and
^{7}Li(n,n'α)T reactions allow a tritium breeding ratio
slightly in excess of unity.   Outside the blanket is a
shield to attenuate the neutron flux and superconduc-
ting magnets which provide the magnetic field which
confines the reacting plasma.   Most of the energy released
in the fusion reactions is deposited in the blanket and
removed by a coolant, which is used with conventional
turbines to generate electricity.

The tokamak is a magnetic configuration which has
been particularly successful in confining a plasma.   The
largest plasma physics experiments now in operation, the
Joint European Torus (JET) at Culham and the Tokamak
Fusion Test Reactor (TFTR) at Princeton, are based on this
configuration.   The tokamak has a toroidal geometry   with
a strong toroidal magnetic field (~ 5 T).   To provide
stabilization of the plasma, a smaller poloidal magnetic
field is also necessary, generated by a current flowing in
the plasma.   Since this current is linked both to the
magnitude of the toroidal magnetic field and the
dimensions of the plasma, it is a good measure of the
confining ability of a tokamak system;   the maximum current in the JET experiment is 7 MA and it is expected that
the current in a tokamak fusion reactor will be of the
order 20 MA.

Surrounding the plasma is the blanket, whose functions are to convert the kinetic energy of the neutrons
into heat and to breed and recover tritium.   The breeder
material is essentially lithium, which may be in either
liquid or solid form.   Liquid breeders can be either pure
lithium or mixtures such as lithium lead.

Solid breeders which have been considered include
lithium oxide, lithium aluminate, or lithium zirconate.
The main requirements for a breeder are that it allows a
tritium breeding ratio greater than unity, that it does
not bind the tritium too tightly and thus give a very high
tritium inventory in the blanket, and that it can operate

at temperatures of 400 to 600°C without serious corrosion. The advantage of liquid breeder materials is that they may be circulated through the blanket and the tritium extraction accomplished in plant external to the reactor, whereas in the case of solid breeders the tritium must be removed by a purge gas within the blanket. A variety of coolants have been considered for the first wall and blanket, of which the main possibilities are pressurized water, helium or carbon dioxide.

The breeder material and coolant must be contained within a mechanical structure, which must withstand mechanical and thermal stresses and must be resistant to radiation damage. In addition to void swelling, which is common in fission reactors, the structural material may also suffer enhanced embrittlement as a result of the high level of n,α reactions due to the high energy of the neutrons. As a result, the blanket structure may have to be replaced several times during the life of the reactor. The neutron induced radioactivity of this structural material is one of the main environmental hazards of a fusion reactor, both because of doses to workers during maintenance and decommissioning activities, but also because of the quantity of radioactive waste material generated. On the other hand a wide range of materials can be considered, and the possibility of developing low activation materials is being pursued.

Reactor Parameters

As the basis of the recent study for the Commission of the environmental safety and economic potential of fusion reactors, a set of reactor parameters was developed representing reactors that might be available around the year 2050.[8] Table 1 shows parameters of three such reactors. The first of these is a prototype commercial sized reactor (PCSR-E), based on present experimental results, and is therefore a very conservative design. The other two are based on modest extrapolations of present results and are therefore more characteristic of what might be possible after the operation of the first generation of experimental reactors (NET or ITER). In all cases the net electrical power output of the station was set at 1200 MWe. The most noticeable change in the parameters of these three reactors is the physical size, as characterized by the plasma major radius, and as this diminishes the neutron wall loading at the first wall increases, giving more compact and therefore more economic designs.

Table 1. Parameters of three possible tokamak reactors
with 1200 MWe net output [Ref 8]

	PCSR-E	R1	R2
Troyon coefficient (Tm/MA)	0.035	0.030	0.040
Plasma elongation	1.7	2.0	2.25
Stress in coils (MPa)	160	200	250
Thermal efficiency (%)	35	35	40
Plasma half-width (m)	2.4	2.0	1.4
Plasma major radius (m)	9.3	7.1	5.3
Plasma current (MA)	16	22	16
Plasma pressure, β	3.8	5.1	7.6
Wall loading (MW/m³)	2.2	3.1	4.5

Of the three reactors shown in Table 1, reactor R2
was taken in the EEF study as the basis of further studies. On the basis of present radiation damage studies,
it is expected that the first wall and blanket structure
will have to be replaced 7 times during the 25 year life
of the reactor. This material will be lower in volume
than the shield, but will make the greatest contribution
to the activity of the waste. The divertor targets will
be replaced even more frequently, but are smaller
components.

3   TRITIUM

Two chemical forms of tritium are of main interest,
elemental tritium (HT) and tritiated water (HTO). Elemental tritium is not easily absorbed by the body and the
most significant route for exposure is irradiation of
lung tissue. Tritiated water, which behaves as normal
water, is absorbed into the body both through lung tissue
after inhalation and by direct absorption through the
skin. Once absorbed into the body, tritiated water is
rapidly distributed and irradiates all soft tissue. The
major fraction has a biological half life of about ten
days, corresponding to the normal water turnover rate. A
small remainder, which is retained for longer periods,
makes an insignificant contribution to the dose. Because
of the relative ease with which the body absorbs water,
tritiated water is more hazardous as an airborne material
than elemental gaseous tritium by a factor of about 25000.
Thus, although tritium is more likely to leak from a

reactor in the elemental form, and is only slowly
converted to tritiated water in the environment, it is
usual to be pessimistic and assume that all tritium
released to the environment is in the form of tritiated
water.

Total tritium inventories in a fusion reactor include the amounts in the reaction zone, the breeder blanket and the tritium processing plant, and may be up to 12
kg (4×10^6 TB_q). The largest and most vulnerable
component is in the blanket, and in various reactor
designs is in the range 100 g to 10 kg. These levels are
comparable with inventories in existing CANDU reactors,
for example 3.5 kg in the four 550 MWe reactors at
Pickering A.

Operational Releases of Tritium

The global effect of continuous tritium releases in a
fusion-powered world economy has been estimated.[1]   If
fusion power provided 6×10^6 MWe (0.08% of solar input)
and each 3 GWe station discharged 30 kCi/y (3 TB_q/day),
the steady state activity of tritium in the environment
would be about 4×10^7 TB_q. Mixed with the atmosphere and
upper layers of sea, this would result in an increased
dose to man of 0.2 mrem/y (2 µSv/y), corresponding to a
worldwide collective dose of 8×10^5 man.rem/y (8×10^3
man.Sv/y) worldwide, which is about 0.04% of the
collective dose from natural background radiation. (In
this review, the units used in the quoted references have
been retained, and their SI equivalents added.)

A more critical factor in deciding upon acceptable
loss rates may be the dose to the most exposed individual. In a recent calculation[9] a fusion reactor was
assumed to release 7.1 kCi/y (260 TB_q/y) of tritium from a
100 m high stack. The maximum dose was received by a
person living at the site boundary at a distance of 320 m
from the reactor in the direction of the prevailing wind,
who was assumed to consume only food grown at the location. The estimated 50 year dose commitment resulting
from one year's exposure was 0.029 mrem (0.3 µSv), which
is a factor 1000 less than the limiting critical group
dose of 1 mSv/y recommended by ICRP and adopted in the UK.

The estimated collective dose to the whole population living within 80 km of the reactor was 3.4 man.rem (0.03 man.Sv), which is 10^{-5} times the collective dose from natural radiation.

A similar calculation assumed the release of 910 Ci/y (34 TB_q/y) of tritium to a local stream. The 50 year dose commitment for a person drinking the water and eating fish from the stream was estimated to be 0.037 mrem (0.4 μSv). If, on the basis of these calculations, it is considered that releases from a fusion reactor of the order of 4 TB_q/day would be acceptable in appropriate locations, this implies a requirement of tritium system leak tightness of the order of 1 ppm/day of the total inventory. For comparison, the loss rate from Pickering A was about 5 TB_q/day over the period 1977 to 1981.

The routes for tritium escape during normal operation include the fuel injection and reprocessing equipment, and coolant pipes and valves, in addition to occasional leakage during maintenance operations. The largest loss is expected to be from the coolant through the heat exchanger to the steam system, and in the Starfire reactor design the loss was estimated to be 0.8 TB_q/day from a water-cooled blanket, but only 0.008 TB_q/day for a helium-cooled blanket. In the absence of practical experience from tritium-fuelled fusion reactors, the first indications of possible tritium loss rates will come from the Tritium Systems Test Assembly at Los Alamos, which is now operating with a tritium throughput of about 1 kg/day and is required not to exceed a loss of 8 TB_q/y under normal operation.

Accidental Tritium Releases

Only part of the tritium inventory is vulnerable to release in an accident. There are several factors which suggest that the effects of such a release will be smaller than the effects of a major fission reactor accident, such as its rapid rate of dispersion and the absence of any known mechanism for its concentration in the body.

The consequences of a major release of tritium have been estimated by Häfele,[3] who assumed that 40% of a total inventory of 25 kg, i.e. 10^8 Ci (4 x 10^6 TB_q) was released in the form of tritiated water. Using a model similar to that used for fission reactor accidents (Rasmussen Report), the initial dose to bone marrow was calculated at

Table 2. Critical doses to bone marrow from accidental
releases of tritium from a fusion reactor and fission
products from a fission reactor [Ref 3]

Distance km	Critical Dose (rem) Fusion	Fission
0.5	2.6 E3	1.3 E5
1	6.9 E2	3.8 E4
2	1.9 E2	1.0 E4
5	3.4 E1	1.9 E3

various distances from the release. Some results are given
in Table 2, compared with the results for a similar calculation for a fission reactor in which 40% of the iodine and
tellurium was assumed to escape. The estimated consequences
of the tritium release are a factor 40 lower than the fission
reactor release.

The above estimate of the consequences of a tritium
release must be considered an extreme example. The assumed
inventory was much larger than is considered acceptable in
most fusion reactor designs, and the mechanisms which would
allow the release of tritium as tritiated water need to be
considered. The more recent ESECOM study[5] concluded that the
active inventories of tritium in fusion reactors are small
enough that even complete release under adverse meteorological conditions would not produce any prompt fatalities off
site. However, as in fission reactors, the probability of
such an accident would have to be demonstrated to be less
than 10^{-6}/year. For comparison the JET experiment will be
allowed to hold a maximum inventory of 95 g of tritium,
distributed between the experiment, the tritium processing
plant, and a store, which is not considered to be a hazard to
the local population.

4   ACTIVATED STRUCTURAL MATERIALS

Although the fuel cycle of a fusion reactor does not involve
the input of radioactive material nor generate radioactive
waste products, the neutron flux from the plasma will generate parasitic radioactivity in the reactor structure. Consequently, fusion reactors will generate radioactive waste
during operation from routine replacement of irradiated
components and following decommissioning. The quantity of
this waste has been estimated[10] by surveying six D-T conceptual reactor designs (3 tokamaks, 2 reversed field pinches

Table 3. Estimated quantities of waste from the
operation and decommissioning of six conceptual fusion
reactors [Ref 10]

	Total Waste		Repository Waste (net)		Repository Waste (packaged)
	Mass te/GWe	Volume m³/GWe	Mass te/GWe	Volume m³/GWe	Volume m³/GWe
STARFIRE	22800	3660	11600	1940	10400
DEMO R254	30700	4830	12400	1990	10700
PCSR-E	42200	9990	28600	8050	43400
CRFPR	8340	906	8340	906	4870
TITAN	4800	1120	4800	1120	6030
WITAMIR	30900	4380	27200	3900	21000

and a tandem mirror). Since operational waste from
routine replacement of the blanket will be stored on site
until decommissioning and disposal, no distinction is made
between the two. As noted earlier, the volumes of the
shield and magnetic field windings are much greater than
the volumes of the nuclear blanket, and therefore the
quantity of waste is not strongly dependent on the load
factor or availability of the reactor.

    The estimated quantities of waste, normalized to
1 GWe full power rating are summarized in Table 3. The
volumes of packaged repository waste vary considerably,
being smallest for the reversed field pinches which are
compact machines and largest for the PCSR-E reactor, which
has the most pessimistic physics assumptions. The three
tokamak reactors produce volumes of waste which differ by
a factor of four. The PCSR-E gives the largest volume for
several reasons, including the fact that all material from
the shield is assumed to be repository waste, whereas for
Starfire and Demo a substantial proportion of the shield
is non-repository waste. This emphasizes the point that
the shield contributes the largest component of the waste
and suggests that the use of low-activation steels and
suitable design may have a substantial influence on the
levels of waste produced.

    A rough estimate of the volume of fully compacted
waste from the core of a fusion reactor can be obtained by
assuming that all components are constructed from steel
and last for the whole life of the reactor. In practice

the actual volumes of waste requiring disposal in a repository fall above or below this line for a number of reasons. The scatter in the more detailed estimates obtained
in both the Culham and ESECOM studies relative to such a
rough estimate suggests that the volume of waste from a
fusion reactor is not known within an accuracy of a factor
3. The results also show a clear trend to lower waste
volumes for designs with a high power density. A power
density of greater than 100 kWe/te has been proposed as a
criterion for the economic viability of a fusion reactor
and, if this is accepted, the volume of unpackaged waste
should not exceed 1000 to 2000 m³/GWe, provided components
with large voids are reasonably compacted and the shield
design is optimized so that a significant proportion does
not require disposal in a repository.

These volumes of waste generated by a fusion reactor are somewhat larger than the volumes generated by a
fission reactor. However, it must be emphasized that the
volume of waste, although influencing the economics of
waste disposal, is not the major factor in estimating the
risks. These are considered in the next section.

Disposal of Activated Material

Several assessments have been undertaken of the consequences of disposing of waste from a fusion reactor[11],
[12],[13] taking into account both the normal evolution and
the possibility of intrusion for shallow and deep repositories. The most recent study[13] was based on two
variants of a reactor with parameters corresponding to
reactor R2 discussed in a previous section. The first of
these used a blanket constructed from conventional steel
and the other a low activation vanadium alloy. The
estimated risks were compared with those from the disposal
of material from a pressurized water reactor (PWR). It
was assumed that the material was placed in a deep geological repository for which the pathway for the escape of
activated material was by ground water. Calculations
focused on this pathway, although brief consideration was
also given to two others, the release of radioactive gas
and human intrusion.

Consideration of the ground water pathway was limited
to the near field of the repository and to the geosphere.
Radionuclide concentrations were calculated in near field
pore water and in the pore water emerging from the geo-
phere into the biosphere. To estimate the overall consequences, a simple biosphere model was used; the   total

flux of radionuclides leaving the geosphere was assumed to enter a typical stream, which in turn was used as the sole drinking water supply for an individual. Calculations for the geosphere were conducted in two stages; the movement of ground water through the repository and geosphere was considered to determine the time taken for water to return from the repository to man's environment, and then the transport of nuclides by the ground water was estimated.

The estimated toxicities are shown in Table 4. The toxicity of the near field pore water after 300 years for the PWR case is three orders of magnitude above the two fusion cases, due to the high inventories of the fission products ^{90}Sr and ^{137}Cs in these wastes. At very long times the toxicity for the two fusion cases decays more rapidly than that for the PWR, because the very long-lived ^{238}U and ^{235}U are only leached from the repository very slowly.

The doses arising from drinking stream water show distinct pulses for all the inventories. The first arrives after 10^5 years (the ground water return time) and arises from the radionuclides that are poorly absorbed on far field materials but are sufficiently long-lived to emerge before they decay to insignificant levels. The most important for the PWR is ^{129}I, whereas for the fusion wastes the main contributions are from ^{36}Cl and ^{10}Be, and are nearly a factor 1000 lower. The maximum dose rate of 1.6×10^{-10} Sv/y for fusion may be compared with the limit for a UK repository of 0.1 mSv/y. At later times there is a second significant pulse for the PWR arising from ^{226}Ra, a short-lived daughter of ^{238}U, which derives from uranium that has migrated most of the way as a long-lived

Table 4. Estimated toxicity resulting from disposal of material from fusion reactors with conventional and low-activation materials and a PWR [Ref 13]

	Near-field Toxicity (Sv/y)		Stream Water Toxicity (Sv/y)
	300y	10^5y	maximum
Fusion 1	6.2 E-1	2.0 E-3	1.6 E-10
Fusion 2	6.3 E-1	1.1 E-3	2.9 E-10
PWR	3.5 E+2	6.1 E-1	9.1 E-8

radionuclide but decayed before emerging into man's environment.

The hazards from waste disposal have also been assessed in the ESECOM study. In this case the wastes from several different fusion reactors and a large scale prototype breeder (LSPB) fission reactor were assumed to be placed in a shallow repository, and various indices calculated on the basis of intrusion after 100 years either to build and live on the site or excavation of the site. Some results are given in Table 5, which shows a range of results for the fusion reactors, but several orders of magnitude advantage for fusion relative to fission. Although this study showed that all but one fusion reactor qualified for shallow land burial according to current American regulations, the authors do not claim that this would necessarily be the most appropriate management method.

Accidental Releases of Activated Materials

The majority of the induced radioactivity in a fusion reactor exists in the structural material of the first wall and blanket. The breeder material also contains induced activity as well as tritium, and there is some induced activity in the inner layers of shielding. At the present state of conceptual designs of fusion reactors it is possible to estimate the quantities of activity and to postulate mechanisms for its escape, but not to be quantitative as to the amounts which may escape to the environment nor as to the frequency or consequences of its escape.

Table 5.   Estimated hazards from the waste from conceptual fusion reactors and a large scale prototype breeder fission reactor [Ref 5]

	Intruder dose rem	Intruder hazard Rm^3/y	Deep disposal index m^3
Tokamak,V,Li	0.22	12	41
Tokamak,RAF,He	0.039	3.2	33
RFP,RAF,PbLi	35	670	40,000
Tokamak,SiC,He	0.004	0.54	0
LSPB (fission products only)	34,000	140,000	8,200,000

To obtain a preliminary estimate of the consequences
of an accident, however, the ESECOM study divided the
material in fusion and fission reactors into five categories and calculated the fractions which would have to be
released to the environment to have a given consequence.
The categories range from elements which are gaseous or
extremely volatile under thermo-chemical conditions
(Category I) to those which are resistant to volatilisation even under extreme conditions (Category V); examples
of materials in Category I are tritium and argon and
materials in Category II rhenium and caesium and in
Category V iron and plutonium. These categories are
based, where possible, on experimental data, and assume
that the elements come into contact with oxygen at high
temperature. Threshold release dose fractions (TRDFs)
were calculated which would result in either a critical
whole body dose commitment of 200 rem (2 Sv) to an individual at 1 km from the accident or a 50 year whole body
dose of 25 mrem (0.25 mSv) at a distance of 10 km. The
former is the smallest whole-body dose with an appreciable
chance of causing an early fatality, and the latter is the
level in the USA above which it would be necessary to initiate special clean-up measures or to relocate the local
population.

Some results of such calculations are given in Table
6. The fraction of the total material in a reactor which
would give the specified consequences are shown for aggregated categories, on the basis that an accident releasing
material in any category would also release the material
in a lower category. Fractions higher than unity represent a significant safety factor since the total inventory
falls short of the quantity of material which must be
released to cause a relevant consequence; small fractions
are most troublesome for the high mobility categories.
Results for fusion reactors are presented separately for
the first wall and the remainder of the reactor, since the
former is the main source of afterheat and suffers the
highest temperatures in a loss of coolant accident.

The results in Table 6 show that for the high mobility materials, fusion reactors have a large safety factor
and that for materials of lower mobility the threshold
release dose fractions are consistent with accidents
localised to one area of the first wall or blanket. The
only accident which is thought capable of causing high
temperature damage to the whole blanket is a fire in a
reactor with liquid lithium breeder, and for this reason
this breeder has not been considered in European conceptual designs.

Table 6.   Estimated Threshold Dose Release Fractions
based on a 25 rem chronic dose limit.   For fusion reactors,
estimates are given separately for the first wall and the
remainder of the reactor [Ref 5]

	I	I-II	I-III	I-IV	I-V
Tokamak,V,Li	1.1E1	7.7E-1	5.2E-1	2.2E-1	1.7E-3
	2.6E2	8.2E-1	1.1E-4	1.0E-4	9.3E-5
Tokamak,RAF,He	2.5E1	2.2	1.1E-4	1.1E-4	1.1E-4
	3.4E3	8.4E-1	1.1E-4	1.0E-4	9.8E-5
Tokamak,SiC,He	2.5E1	2.5E1	7.0	2.8	2.8
	7.7E1	1.2E1	9.5E-2	1.6E-2	1.6E-2
PWR	2.8E1	1.3E-4	1.2E-4	8.6E-5	4.8E-5
LSPB	6.3E1	1.7E-4	1.6E-4	9.5E-55	4.2E-5

     Possible designs for the first wall and blanket of a
fusion reactor have been considered, involving a range of
materials.   Levels of afterheat have been calculated and
the consequences for the blanket of a loss of flow or loss
of coolant accident estimated, although without the inclusion   of chemical reactions or fires.   Temperatures in
the blanket were estimated in the ESECOM study on the
assumption that the first wall neutron loading was 5 MW/m^2
which is roughly the level required for economic power
production in a tokamak reactor.   Peak temperatures in the
first wall range from 800 to 1200°C, reached in a few
hours.   These temperatures may be sufficient to cause
permanent damage to the first wall and blanket structure,
but not to cause an extensive release of active materials
other than the most mobile.

     Studies of the extent to which radioactive material
released in accidents would be contained within the reactor structure or building are still at a very early
stage.

Low Activation Materials

     The neutron activation of reactor materials has
implications affecting every aspect of the reactor system,
including the consequences for accidents and the need to
dispose of radioactive waste as discussed in the previous
sections.   However, since the induced radioactivity is not
intrinsic to the fusion process itself but depends on the
materials surrounding the reacting plasma, an opportunity
exists of increasing the environmental advantage of fusion
through the careful selection of materials.   The use of
low activation materials has mainly been considered for

the first wall and blanket structure, since these
represent the highest proportion of the radioactive
inventory in a reactor, but the use of low activation
steels in the shield might allow recycling of these
materials to significantly reduce the quantities of
radioactive waste.

In the short term, most effort has been directed at
the development of steels with alternative alloying
elements but with mechanical properties comparable to
those of traditional steels. For example, nickel in
conventional austenitic steel may be replaced by manganese
and nitrogen. Immediately after discharge from the
reactor the activity of such steels is similar to the
activity of conventional steels, but after storage for 100
years their activity may be lower by more than an order of
magnitude. Thus low activation steels do not significantly affect the hazards from accidents, but offer
advantages in waste disposal.

Greater improvements can be achieved by the use of
vanadium alloys and, with these materials, the induced
activity on removal from a reactor may be reduced by a
factor 10 and after 100 years may be reduced by more than
factor 1000. Vanadium alloys are in general resistant to
radiation damage, but a substantial programme of development would be necessary to produce materials suitable for
use in the fusion environment and with mechanical and corrosion properties comparable with those of existing
materials. Furthermore, to achieve the potential advantages of vanadium alloys it would be necessary to produce
materials with comparatively low impurity content.

As an illustration of the environmental advantages of
low activation materials, Table 7 shows the estimated
maximum individual doses and the risks associated with the
burial in a deep repository of a conventional and a low
activation martensitic steel and a vanadium alloy from the
structure of a fusion reactor.[12] It is seen that there
are modest gains from the use of a low activation steel,
and greater gains from the use of a vanadium alloy. In
the former case the majority of the dose is due to alloying elements (0.02% of nitrogen and 0.7% of tungsten),
whereas impurities begin to become significant in the
vanadium alloy (4 ppm niobium).

It should be noted that the results quoted in the
previous paragraph are significantly different from the

Table 7.   Estimated radiological impact associated with
the burial of 10^5 tonnes of various structural materials
from a fusion reactor in a deep repository   [Ref 12]

	FV448	LA12TaLC	V-3Ti-Si
Normal evolution			
Max. individual dose Sv/y	3.1 E-6	2.3 E-7	2.7 E-9
Total risk y^{-1}	3.9 E-8	2.9 E-9	3.4 E-11
Borehole intrusion			
Max. individual dose Sv/y	2.4 E-1	2.4 E-4	3.7 E-4
Total risk y^{-1}	1.5 E-7	1.4 E-10	2.3 E-10

results of an earlier similar study.[11]   This was partly
because the compositions of the low activation alloys in
the earlier study were optimized to minimize their contact
dose rates, which is relevant to reactor maintenance.
This emphasizes the fact that low activation materials
can be developed according to several different criteria,
and do not necessarily give similar advantages in all
situations.   The difference was also partly due to recent
improvements in the accuracy of the nuclear data required
for the environmental assessments.   For these and other
reasons, the development of low activation alloys for use
in fusion reactors is still at an early stage.

Recycling of Low Activation Materials

The development of low activation materials opens the
possibility of an alternative strategy for waste manage
ment.   If the contact dose rates from structural materials
fall to an acceptable level after periods of about 50
years, it may be possible to re-use the materials, fabri-
cating new reactor components in lightly shielded
facilities.

The radiological and economic implications of recycling structural materials has been considered in the case
of low activation steels.[14]   For the recycled material it
was assumed that they passed through five cycles before
eventual disposal.   The two alternative waste management
routes were defined and the plant and equipment requirements for initial processing, storage, packaging, transport, disposal, remelting, conversion and fabrication and
reassembly established.   Costings took account of the
capital cost of plant, equipment and raw materials as well
as operating, maintenance and labour costs.   Occupational

radiation exposures were calculated for normal operation and maintenance tasks from estimates of manning levels and area dose rates. Doses to members of the public were calculated for exposure to direct radiation, airborne releases and liquid effluent releases and the risks from post-disposal intrusion were estimated.

The estimated total annual occupational radiation doses for recycling averaged over five cycles was approximately 50% greater than that for direct disposal. The total exposure was dominated for direct disposal by the dose from initial processing operations at each reactor site, and for recycling by initial processing and reassembly. Doses to members of the public from processing, storage or transportation were comparable for the two waste management options; in the long term, doses to the public would be dominated by the final disposal. The overall cost of recycling was estimated to be 35% less than for direct disposal. This assumed, however, that the cost of replacement low activation steels was a factor 2.3 higher than the cost of conventional steels. If, in the future, low activation steels were to cost the same as conventional steels, the overall costs of recycling and direct disposal would be similar. The similarity in costs of the two options, together with the slightly lower occupational dose and short term public doses associated with the direct disposal route indicate that initially direct disposal is the preferred waste management option. However, other factors may eventually change this judgement, and in particular changes in public attitudes to the use of material resources and the quantities of radioactive waste sent for disposal.

5 CONCLUSIONS

A number of studies, based on relatively simple conceptual design studies of fusion reactors, suggest that fusion reactors appear to have an inherently lower environmental impact than fission reactors - although the realization of this potential has yet to be demonstrated.

One of the major hazards to the public from a D-T fusion reactor will be associated with its tritium inventory. To restrict discharges to the environment to levels of the order of 4 TB_q/day, a high standard of containment will be required, comparable to that achieved in CANDU reactors.

The effect of a severe accident leading to the
release of a substantial fraction of the tritium inventory, however, will be less than the effects of a comparable fission reactor accident. This is because tritium
disperses more rapidly, with a biological half-life in
soil or humans of about 10 days.

The other major environmental impact of fusion is
from activated constructional materials. The volume of
waste from decommissioning of a fusion reactor is larger
than from a fission reactor, but the total risk after
disposal in a repository is substantially lower.

A further advantage for fusion is that induced
activity in the constructional materials is not intrinsic to the fusion process, so that a wide choice of
materials is possible. Low activation materials may
therefore be developed to further reduce the consequences of accidents or waste disposal. The prospect also
exists of recycling low activation materials to reduce
waste quantities and conserve scarce resources.

In this review the environmental impact of fusion
reactors has been compared with the impact of fission
reactors, since they are both nuclear sources of energy;
the comparison is not intended to claim advantages for one
system or the other. It is accepted that the development
of fusion power is still at an early stage, and that more
detailed fusion reactor designs are required before their
environmental consequences can be predicted with any
accuracy. Never-the-less, fusion reactors appear to offer
sufficient environmental advantages to justify the
development of this major alternative source of energy for
the next century.

6   ACKNOWLEDGEMENTS

The author acknowledges the contribution of many
colleagues to the environmental studies quoted in this
review.

7 REFERENCES

1.   R Carruthers et al, Culham Study Group report on
     fusion reactors and the environment, CLM-R148,
     June 1975.

2.   F N Flakus, <u>Atomic Energy Review</u>, 1975, <u>13</u>, 588.

3.   W Häfele et al, Fusion and fast breeder reactors,
     RR-77-8, Revised July 1977

4.   R Hancox and W Redpath, Fusion reactors – Safety
     and environmental impact, CLM-P750, May 1985

5.   J P Holdren et al, Report of the Senior Committee
     on environmental safety and economic aspects of
     magnetic fusion energy, UCRL-53766, September 1989.

6.   Environmental impact and economic prospects of
     nuclear Fusion, EURFU BRU/XII-828/86, November 1986.

7.   R S Pease et al, Environmental, safety related and
     economic potential of fusion power, PEC-ED/050,
     December 1989

8.   P I H Cooke et al, Parameters of a reference tokamak
     reactor, CLM-R298, December 1989

9.   C Easterly et al, <u>Fusion Technology,</u> 1989, <u>16</u>, 125

10. J D Jukes et al, Waste generated by fusion reactors.
     To be published.

11. J P Davis and G M Smith, Radiological aspects of the
     management of solid wastes from the operation of D-T
     fusion reactors, NRPB-R210, November 1987.

12. K R Smith and G J Butterworth, The radiological
     impact of fusion waste disposal, IAEA Technical
     Meeting on fusion reactor safety, Jackson, April
     1989.

13. M G Sowerby and R A Forrest, A study of the
     environmental impact of fusion. AERE R 13708,
     April 1990.

14. M J Plews et al, The cost benefit of recycling low
     activation steels from fusion reactors, CLM-R296,
     August 1989.

The Disposal of Solid and Liquid Wastes from Coal Washing Operations

D. W. Brown

BRITISH COAL CORPORATION, ASHBY ROAD, STANHOPE BRETBY, BURTON-ON-TRENT, STAFFORDSHIRE DE15 0QD, UK

1. INTRODUCTION

The advent of mechanised coal mining techniques has not only increased productivity but also the quantity of dirt mined and brought to the surface. Hard information on the amount of spoil for disposal is not available but estimates derived from deep-mined coal production indicate that the amount generated in 1988 was about 45 million tonnes. Although steps are being taken to reduce the quantity of dirt cut at source with the coal; ie at the coal face, it will still be necessary to clean or blend the output from a mine (termed run-of-mine coal) to produce a high quality and consistent fuel for today's markets. It is necessary for the coal to be 'washed' and the subsequent products (both coal and dirt) de-watered prior to sale or disposal. Since all of the coal preparation plants have closed water circuits it is important that as much water as possible is recovered from the coal and dirt prior to sale or disposal respectively, for re-use and to prevent the discharge of dirty effluents in to rivers etc.

The coarser sizes, for example greater than 0.5 mm in size are relatively easy to de-water. The dirt (and coal) passes from the coal cleaning process onto vibrating screens fitted with wedge wire screen decks where it first is washed using water to remove any contaminates (eg magnetite, from the dense medium baths), which can then be recovered and re-used. As the dirt passes further down the screen most of the surface water is then shaken from the solids until they are discharged from the screen having a moisture

Table 1 : Typical Analyses

	Flotation Tailings	Classified Cyclone Overflow
Solids Content range (% w/w)	30 - 45	30 - 38
Solids Relative Density (approximately)	2.40	1.90
Solids Ash Content (% w/w dry basis)	60 - 85	55 - 70

Solids Size Distribution

+ 1000	0.07	NIL
1000 - 500	0.42	0.3
500 - 250	1.80	0.9
250 - 125	5.35	2.1
125 - 63	7.10	5.6
63 - 37.5	4.23	8.0
37.5 - 28.1	1.12	4.2
28.1 - 21.5	8.54	5.3
21.5 - 16.7	15.73	11.4
16.7 - 10.1	10.39	27.9
10.1 - 4.8	18.91	27.4
- 4.8	26.34	6.9

content of about 4-5% by weight depending on particle size. It is then taken for disposal on the colliery spoil heap or landfill site where it is layered and compacted for safe and efficient disposal. Further de-watering takes place through natural evaporation.

It is, however, the very fine sizes, and particularly flotation tailings with a particle size less than 0.5 mm, which are the most difficult and most costly to treat and in today's competitive market represent an added burden on the selling price of coal. Treatment costs for de-watering dirt are not recoverable and must be kept to the minimum to avoid making the product uncompetitive.

Tailings are the waste product from the flotation cells used for fine coal cleaning and are essentially clay with the majority of the solids particles having a size below 63 micrometres. An analysis range is given in Table I. Recently classifying cyclones have replaced froth flotation cells for cleaning fine coal. Since the

particles with size less than 63 micrometres are
predominantly clay (or dirt) classifying the fine solids
suspension in cyclones with a cut point at about 63
micrometres will essentially separate the coal from the
waste. Cyclones are simpler with no moving parts,
maintenance free and more economical than flotation
cells. In addition no consumables (flotation oil) are
required. The overflow from the classifying cyclones,
which is waste, is finer than flotation tailings but
nevertheless is treated in the same way. Although in
this paper only reference will be made to tailings the
same procedure will apply equally to the classified
cyclone overflow. An analysis range of classified
cyclone overflow is given in Table I. The amount of
tailings/classified cyclone overflow currently being
produced, and considered as waste, is estimated at 6
million tonnes per annum.

2. PRIMARY THICKENING

The produced tailings are typically low solids
concentration (3-10% w/v) and high volume (up to 1400
m^3/h), and must be concentrated for economic and safe
disposal. The most economic way of increasing the
solids concentration and recovering the bulk of the
water in a cleaned condition is by the use of a bowl
type clarifier or conventional thickener. The tailings
suspension is treated with a polyacrylamide based
flocculant which then forms bridges between the
individual solid particles to form a comparatively
large, fluffy flocculated mass (known as a floc) which
settles rapidly in the thickener, leaving clear water
overflowing the rim of the thickener for re-use in the
coal preparation plant.

The addition of flocculant must be carefully
controlled, not only to ensure economic utilisation but
also consistent and correct operation of the thickener.
Since both the tailings suspension flowrate to the
thickener and its solids concentration are constantly
varying, addition of flocculant solution at a constant
rate will not produce consistent operating conditions
within the thickener. Overdosing will not only increase
settling rate but will also increase the viscosity of
the settled solids whilst underdosing reduces settling
rate allowing solids to overflow the rim in the
'cleaned' water. Two methods of automatic control have
been used by British Coal. The first, the Clarometer,
is a device which samples the flocculated solids from

the centre curtain of the thickener and measures the
solids settling rate. The measured value is then
compared with the required settling rate and the
Clarometer adjusts the flocculant addition rate
accordingly. If the settling rate is too slow the
flocculant addition rate is increased, and it is
decreased if the settling rate is too quick. The second
technique adds flocculant on the basis of the solids
mass flow rate into the thickener. The equipment used
is a magnetic flux flowmeter, nucleonic density gauge
and a controller with a proportionality setting. It is
assumed that for normal operation of the coal
preparation plant the ash content of the tailings, and
therefore the solids relative density, will vary very
little and hence the solids content of the suspension
can be assumed to be directly proportional to the
relative density of the tailings suspension. The signals
from the flowmeter and density gauge, multiplied
together, give the solids mass feed rate to the
thickener and the proportionality controller is then
used to adjust the flocculant dosage to give the
required settling rate. Once set, the flocculant
addition rate will change with both suspension feed rate
and relative density to maintain the consistent settling
rate.

However, conventional thickeners are usually high
in capital cost and occupy a large area of ground space.
Alternative thickeners evaluated by British Coal, for
example the high capacity and lamella types, are much
smaller in size and lower in capital costs compared to
similar capacity conventional thickeners. Nevertheless
their operation does incur additional running costs,
primarily through increased flocculant dosages of up to
six times that for conventional thickeners, principally
because of the need to rapidly effect solid-liquid
separation due to the smaller volumetric capacity of
these particular thickeners. This smaller capacity, in
addition, renders the thickener more susceptible to
fluctuations in feedrate and solids concentration
thereby making control of operation more difficult.
Furthermore the smaller thickener, unlike the
conventional type, cannot be used as a "buffer storage"
to enable washing operations to continue in the event of
breakdown of subsequent tailings de-watering equipment.

Whatever the choice of thickener it is imperative
that the solids are extracted from the base of the
thickener with as high solids content as possible. For
British Coal concentrations in the range 30-45% are the

Figure 1. A Typical Lagoon

norm and such suspensions are fluid in nature and can be
readily   pumped   from   one   point to another for further
treatment and/or disposal.   Extraction by variable speed
pumps   with   monitoring   by a density gauge to match the
solids   feed   rate   ensures   that   a   high   solids
concentration   is   maintained   to   reduce the loading on
subsequent de-watering process.

3. LAGOONS

     British Coal's preferred and most economical method
of   disposal   of the concentrated tailings suspension is
to pump the thickener underflow directly to a lagoon, or
slurry pond (Figure 1), situated on the spoil heap.   The
lagoons   are   specially   constructed   so   as   to operate
efficiently   and be environmentally acceptable, and care
is   taken   to ensure that there is no escape of effluent
to   streams   and   rivers.   The tailings suspension in the
lagoon   is   allowed   to   concentrate   by   gravity
sedimentation   of   the   solids   over   a   period of time.
Clean water may then be returned to the coal preparation
plant   for   re-use.   When the lagoon is full the dried
solids may be dug out and transferred to another part on
the   colliery   spoil   heap   for   final   disposal.
Alternatively,   another   lagoon   may   be   constructed
elsewhere   and   the full lagoon covered over with coarse
discard   and   soil   reclaiming   the land at the earliest
opportunity   for   other   uses   such   as   agriculture   or
grassland.   Typically, costs for tailings treatment by

Table 2 : Capacity of Filter Presses

FILTER PLATE SIZE (m x m)	FILTER PLATE RECESS DEPTH (mm)	MAX NUMBER OF CHAMBERS	APPROXIMATE DRY SOLIDS THROUGHPUT /CYCLE (tonnes)
0.84 x 0.84	32	40	1.0
1.0 x 1.0	32	80	2.7
1.3 x 1.3	32	120	6.7
1.5 x 1.5	32	120	9.0
2.0 x 1.5	32	150	15.0
2.0 x 2.0	32	160	21.3
2.0 x 2.0	40	150	25.0

this method, including primary concentration in a conventional thickener, are approximately £1.20 per tonne dry solids.

However, the use of lagoons is not universal throughout British Coal. Where space is not available and/or there are difficult geological conditions the thickened tailings suspension at a solids concentration of 30-45% by weight must be further de-watered to obtain a drier material which can then be blended with the coarse discard from the coal preparation plant prior to disposal. These further de-watering processes are filter presses, multi-roller filterbelt presses and solid bowl centrifuges.

4. FILTER PRESSES

Filter presses are the longest established of all types of mechanical equipment used by British Coal for the further de-watering of tailings suspensions from thickener underflows. Although filter presses have been in use in a variety of industries it was not until 1984 that the first, a 40 chamber, 0.84 m x 0.84 m unit was installed at a British Coal site at Manvers Colliery. In spite of its longevity in use most of the significant advances in filter presses have taken place within the last 20 years : larger presses, 'plastic filter plates', automation and filter clothwashing/cake discharge.

By the use of the large 2 m x 2 m presses currently available today (Table 2) nearly seven times more suspension can be processed in a single cycle than from a 1.3 m x 1.3 m unit. Furthermore since manpower levels are usually based on the number of machines installed fewer machines mean fewer men. Spares, stocks and

Figure 2. Side Bar Filter Press Fitted with Cake Release
          Device

maintenance costs are also lower when fewer units are
installed. An important technical advantage is that the
heavier weight of the cake produced materially assists
in the discharge of the cake, which is traditionally the
most manpower intensive part of the filter press
operation.

In spite of the development of Monofilament 'Rilsan'
(Nylon 11) filter cloths with special calendering, the
pressed cakes discharge freely, or with minimal manual
assistance, only when the filter cloths are new or have
recently been efficiently washed; even for 2 m x 2 m
filter cakes 40 mm thick and each weighing approximately
220 kg! With successive filtration cycles, without
cloth washing, the filter cloths become progressively
more blinded (dirtier) resulting in an increase in the
number of cakes holding up in the press. Furthermore,
the amount of manual effort required to assist in their
release also increases. It is, therefore, no surprise
that a large number of the filter press developments in
the last 10 years have been for devices to discharge
pressed cakes automatically from a filter press. One
such device is that developed by British Coal for a side
bar filter press (Figure 2). It has now successfully
operated continuously for over two years without failure

of any component at Bolsover Colliery in the Central Area of British Coal. In spite of the value of such a device, filter cloths still became dirty resulting in cake hold-up. Therefore, in parallel with the development of the cake release device, an automatic cloth washing machine was designed to work in conjunction with it. At one time early in 1989 under the most severe operating conditions at Bolsover Colliery, in order to maintain good cake release the filter cloths on one press were washed on a daily basis, i.e. approximately every 10 cycles. The gains in solids throughput rate, through faster filtration and cake discharge times, more than offset the loss of press operating time during the automatic cloth washing period.

The last decade has seen a rapid development in the degree and sophistication of computer and transducer technology. British Coal has taken advantage of this and particularly in the case of filter presses where control of operation has generally been completely manual. Not only can the computer be used to perform all the sequential operations of the filter and its ancillary equipment but through the monitoring of strategically placed transducers, for example such as pressure switches, flow meters etc, the control system can assess the nature of the feed suspension (solids content, solids size distribution etc) and its filterability. Therefore the changeover from fast filling, with a centrifugal pump, to a pressurisation pump can be made when the delivery of the former has decreased to the maximum delivery rate of the latter. Also the filling of the press can be terminated so as to produce a consistent quality of cake (cake moisture within $\pm1\%$ absolute) from one cycle to another. Using a fully automatic control system ensures that the press operations have been optimised and that a high utilisation (of up to 90%) is achieved.

Where these innovations have been installed at Bolsover Colliery, for example, the savings through the reduction of the number of operating presses required from 3 to 2 and the number of men from 6 to 3 (2 to 1 per shift), has almost repaid the initial outlay of £180,000 total development costs within 12 months. In addition the heavy manual effort required from the operators has been substantially reduced.

Another development within the last ten years has been the conversion of a conventional recessed-plate

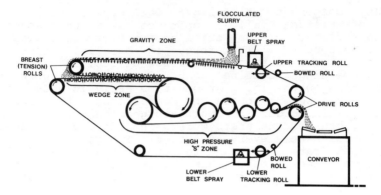

Figure 3. Line Diagram of a Multi-roller Filterbelt
        Press

filter presses to include membrane plates. Membranes
allow additional pressure to be applied directly to each
cake after formation in the press via a rubber diaphragm
or membrane, and as a result eliminate the need to
extend the pressurisation part of the cycle to obtain a
product of acceptable moisture content. Testwork has
shown that the application of membrane plates can either
increase the throughput rate for a given cake moisture,
by about 15% with a possible reduction in costs by up to
£1.00/t dry solids treated, or produce cakes with a
moisture content 2-3% (absolute) lower than that
obtained from a conventional pressure filter at the same
pressure.

     Typically the cost of processing tailings including
primary concentration in a conventional thickener,
depending on the size of press used and on the fineness
of the feed solids (and hence length of filtration
cycle), is £4.50 to £6.50 per tonne dry solids.

     However, since the operation of filter presses
remains a batch operation requiring regular manpower
involvement, consideration was given by British Coal to
the evaluation and use of other continuous equipment
such as Multi-roller Filterbelt presses and Solid Bowl
Centrifuges for the de-watering of flotation tailings.

Figure 4. A Multi-roller Filterbelt Press

5. MULTI-ROLLER FILTERBELT PRESSES (MRF's)

     The advent of high molecular weight polyelectrolyte flocculants gave rise to the formation of very large and stable flocs which had hitherto been impossible with natural flocculating reagents.   Heavy dosing of a tailings suspension containing 30-45% solids by weight with polyacrylamide based flocculants produces a very open and stable structure from which water can escape under gravity.   By the application of low pressure, compared to filter presses, to the floc structure contained between two filter media there is the possibility of gently squeezing further quantities of water from the flocculated suspension.   One such item of equipment developed for this purpose, about 35 years ago, is the Multi-roller Filterbelt Press (MRF).

     MRF's are manufactured by a number of companies; all but a few use the same basic design (Figures 3 and 4).   These consist of two endless belts which pass through a series of pressure rollers of successively decreasing diameter.   As the flocculated suspension passes through the MRF the diminishing radii of the rollers progressively generate increasing pressure and change of point of application (shear) on to the material being de-watered.   However, the underflow from a conventional thickener although slightly flocculated does not have a large, open floc structure to allow free passage for the water during squeezing. To achieve this requires large dosages of, usually, both anionic and cationic flocculant.

Since MRF presses have been used successfully in the paper industry and sewage plants for a number of years, and are also used by private contractors for the de-watering of fines produced in tip washing, a number have been installed in British Coal at Deep Mine sites for evaluation. These machines are 70% lower in capital costs, installation and manpower requirements than filter presses, but running costs (particularly for flocculant - the filter press uses none) are higher. Also, whereas the operation of a filter press is batchwise, MRF's de-water suspensions on a continuous basis.

Evaluation of the MRF's installed at British Coal's deep mine sites has shown for tailings that total flocculant (anionic and cationic) consumption can be up to 1.0 kg/tonne dry solids depending on the nature of solids being filtered, the design and operation of the MRF, and the type of filter media used. The de-watered product, at 30-40% moisture content, by weight, is considerably wetter than that produced by filter presses as the pressure exerted on the tailings by the MRF is both less than and substantially shorter in duration than is achievable in a conventional filter press. (The residence time in an MRF is typically 2-4 minutes compared with a filter press's total cycle time of 1-3 hours). Since the product is wetter than from filter presses, compared to normal plant discard, separate arrangements are made for its disposal : it is allowed to dry prior to blending with coarse discard or alternatively it must be immediately stabilised with cement.

The cost of tailings disposal via primary de-watering in conventional thickeners and using MRF presses with cement stabilisation is in the range £6.00 to £9.15/tonne dry solids. Although the capital cost for such an installation is approximately 40% of that for an equivalent filter press system, running costs in the form of consumables, and particularly flocculant and cement (if used), are substantially higher. Such equipment, however, offers low capital cost options for supplementing an existing tailings de-watering plant where only a small increase in

Figure 5. A Solid Bowl Centrifuge Plant

capacity is required or the additional capacity is required for a short period of time. Furthermore, compared to filter presses, MRF presses can be readily and economically transferred from one site to another, and be installed to full operational order within a very short period of time.

6. SOLID BOWL CENTRIFUGES (SBC's)

Although centrifuges are used throughout the world in a variety of industries and a number of applications to separate solids from liquids, it was less than 10 years ago that the first solid bowl centrifuge (SBC) plant was installed at a Colliery in British Coal's North West Group to de-water tailings at a maximum rate of 75 dry tonnes per hour (Figure 5). The installation followed extensive trials on a single full size centrifuge de-watering underflow from a conventional thickener at a solids concentration of 30-40% solids by weight.

The SBC's, like MRF presses, operate continuously and in order to separate efficiently the very fine tailings particles from the water, the feed must also be pre-treated with polyacrylamide type flocculant. As the feed is subjected to greater shear, than with MRF's,

due to the greater G-forces within the centrifuge this
has necessitated the development of special
polyacrylamide flocculants to give shear resistant flocs
to enable the full benefits of the SBC to be exploited.
The required flocculant dosage was found to be in the
range 0.2 to 0.4 kg/t dry solids treated. The product
from the centrifuges contains only 58 to 64% solids by
weight and has to be stabilised with cement (1.75% to
wet weight of centrifuge product) for safe disposal.

Solid bowl centrifuge installations are 60% lower
in capital than filter presses but running costs are
higher, principally the consumables of flocculant and
cement. Early information shows the cost of tailings
de-watering using conventional thickeners and solid bowl
centrifuges to be approximately £5.00 to £6.00/t dry
solids. However, because the use of this equipment is
relatively new to British Coal long term maintenance
costs are not yet known.

7. DISPOSAL OF DE-WATERED PRODUCT

The de-watered product, whether from filter
presses, MRF's or solid bowl centrifuges, is
non-hazardous and all can be blended with coarse discard
and layered on the colliery spoil heap or disposal site,
or can be used to infill a disused quarry. Strict guide
lines exist not only to control the method of disposal
but also the restoration of the land for re-use.
British Coal co-operates and liaises with councils in
local projects not only for safe disposal of the waste
but for full restoration of the land after completion of
the project, which may also include reclamation of land
from a previous use[1].

8. OTHER METHODS OF DISPOSAL OF WASTE SOLIDS

All the processes considered to date look for the
disposal of the product, from the de-watering process,
to be on the Colliery spoil heap or on land fill/
reclamation sites. British Coal, through the Research
Division of its Headquarters Technical Department, is
investigating the feasibility of disposal of both the
tailings and coarse discard underground; infilling in
the voidage left by the extracted coal behind the face.
This not only has the advantage that the quantity of
waste for disposal on the surface would be reduced but
also that subsidence would be reduced. The project is

in its early stages and it will be about five years
before the technique has been perfected and the
results/economics known.

9. CONCLUSIONS

In the cleaning of the run-of-mine coal to produce
a fuel of consistent and high quality for a discerning
market, there will always be a need to de-water
suspensions of tailings or classified slurries prior to
disposal of the fine waste solids. Since the treatment
of these wastes are a cost to the coal producer and do
not generate an income, because of stringent market
forces they must be safely disposed of by the most
economic means whilst at the same time after final
working of the site leave an environmental acceptable
situation whereby the spoil heap can be developed for
agriculture, industrial development or parkland. Of the
disposal points available British Coal have been able to
improve the environment by providing non-hazardous waste
of coarse discard and de-watered fines suspension for
the infilling of quarries and in the restoration of
land for useful purposes.

Of the techniques currently available to British
Coal the preferred and most economical solution is for
primary de-watering in a conventional thickener followed
by final concentration in a lagoon. Where a mechanical
final de-watering technique is used because of
unavailability of land for a lagoon or because of local
conditions, the choice will depend on the capital
available, the quality of the product required and the
location of the disposal point. Not withstanding this
the oldest established of the mechanical de-watering
devices available, filter presses, still produces the
second lowest (after lagoons) overall treatment costs.

The evaluation of disposal of waste material
underground is continuing but it will probably be about
five years before the technique has been perfected and
the results/economics known.

10. ACKNOWLEDGEMENT

The author wishes to thank Mr C T Massey, Head of
British Coal's Headquarters Technical Department, for
permission to publish this paper and that the views
expressed are those of the author and not necessarily

those of British Coal.

11. REFERENCE

1. C. McDONALD, J.R. RAMSKILL, D.G. CHEESEMAN, and
   D.A. CLARK, 'Welbeck Reclamation and Landfill
   Project : Towards Integrated Resource Management.'
   Mineral Extraction, Utilisation and the Surface
   Environment Symposium, Institution of Mining
   Engineers, Harrogate 1988.

The Disposal of Solid Combustion Products from Power Stations

B. H. M. Billinge*, A. F. Dillon, and D. Tidy
NATIONAL POWER TECHNOLOGY AND ENVIRONMENTAL CENTRE, KELVIN
AVENUE, LEATHERHEAD, SURREY KT22 7SE, UK

F. Harrison
NATIONAL ASH, NATIONAL POWER, TREGONIS HOUSE, SWINDON, UK

1 INTRODUCTION

Products Arising Directly from Combustion

About 80% of the electricity produced in England
and Wales is generated in fossil fuel fired power
stations of which hard coal makes up 94%. In recent
times this has amounted to more than 70 Mte of coal per
year and with an average ash content of 17%, has
resulted in the production of ∿12-13 Mte of ash per
year. This ash may be classified into three types,
furnace bottom ash (FBA ∿ 2 Mte p.a.) pulverised fuel
ash (PFA) and cenospheres (which make up ∿1% of the
whole). The disposal of such a large quantity of
material represents a major challenge which has
stimulated the development of uses for this material as
opposed to disposal as landfill.

Outlets for ash which can make significant inroads
into the large quantities generated are all virtually in
civil engineering and building construction, where it is
generally in competition with other low-cost industrial
by-products or quarried alternatives. Responsibility
for marketing ash for National Power rest with National
Ash.

Research into possible uses of the material began
before the Second World War but it was the rapid
increase of coal-burning for electricity production
after the war that prompted large-scale use of PFA,
particularly in the USA where the US Bureau of
Reclamation published data on its use in the
construction of the Hungry Horse Dam (1). By the middle

sixties National Standards for PFA were established in
the USA, Japan, Great Britain, France and the USSR.

FBA is a coarse dark glassy material recovered from
the high temperature regions of the boiler. It is all
used in block making and road foundation work. PFA, on
the other hand, still needs further markets. It is a
fine dust containing mainly spherical particles
(1-100 μm diam.). There is wide heterogeneity particle
to particle but is basically a fired aluminosilicate
clay containing 60-90% glass. It is the glass content,
in conjunction with the fineness, that gives it one of
its more important properties namely, its pozzolanic
nature. It will therefore react with lime in the
presence of water at ambient temperatures to form
insoluble, cementitious products. As a consequence it
is now used in great quantity to produce building
blocks, civil engineering fill, lightweight aggregate,
cement replacement material, grouting, brickmaking etc.,
which will be explored further in this paper.

Products Arising Indirectly

In more recent years a further facet of the
disposal of solid combustion products has been brought
into focus by the requirement to reduce sulphur and
nitrogen oxide emissions to the atmosphere. Fossil
fuels contain sulphur which is released mainly as
sulphur dioxide on combustion. There have been at least
a hundred methods proposed for its removal from flue
gases (2) but only a small number have been developed to
the full scale. The majority of these utilize limestone
or lime as the alkali medium for fixing the sulphur
because of its large-scale availability and comparative
inexpensiveness.

Direct injection methods e.g. of powdered limestone
into the combustion chamber or sprayed in as a lime
slurry further down the flue gas train both result in
the trapping (in the ESPs or bag filters) of a
complex mixture of lime, calcium sulphite and sulphate
with PFA which can create disposal problems. More
favoured methods are those that produce a marketable
material and in Japan, West Germany and the USA the
limestone/gypsum process, in which sulphur dioxide is
reacted in a limestone slurry absorber under oxidising
conditions to produce gypsum, is the favoured process.
This process has also been adopted in this country and
construction of an FGD plant has commenced at the
4000 MW station at Drax, Yorkshire. Although gypsum of

the right quality is a saleable material, production
from this installation alone is expected to account for
∿30% of the UK gypsum market, so contingency plans for
other methods of disposal must be considered.
Furthermore, the reduction of the emission of nitrogen
oxides is also the subject of international targets and,
depending on which method is adopted can result in
increased carbon content of the PFA or its contamination
with ammonia. These will also be discussed later.

2 COAL ASH

Burning finely powdered coal results in an ash residue
of which 80% is pulverised fuel ash (PFA) which is
carried through the combustion train and removed from
the flue gases by electrostatic precipitators. It is
then either (i) transported by air or mechanical
conveyance to dry storage bunkers (ii) 'conditioned' by
mixing with water and stockpiled until required or (iii)
transported hydraulically to 'lagoons' (from which it
can also be recovered subsequently). The remaining 20%
of the ash, 'Furnace Bottom Ash' (FBA) results from
agglomeration of ash in the high temperature regions of
the plant from whence it is removed to a crusher and
storage. This material is chemically similar to PFA but
its coarser structure makes it particularly suitable for
building block production.

PFA as a Cementitious Material

Chemically PFA is mainly an aluminosilicate glass
material containing variable quantities of iron, sodium,
magnesium, calcium and titanium within the glassy
matrix; the major crystalline constituents are mullite
($3Al_2O_3.2SiO_2$), quartz (SiO_2), iron oxides (haematite
and magnetite) and also some unburned carbon. Table 1
shows the range of UK fly-ash analyses, with carbon from
large base-load stations tending to the lower figure.

Pozzolanic activity is the reaction of the alumino
silicate glassy phase with the lime liberated during the
hydration of Portland cement. The pozzolanic reactions
are not fully understood but are enhanced by high glass
content and greater fineness. The cementitious
products, calcium silicate and aluminate hydrates, are
similar to those from ordinary Portland cement (OPC)
although with PFA/OPC mixes the CaO/SiO_2 ratio tends to
be lower. The reaction of lime with the PFA results in
reduced permeability since the lime would otherwise - as

in an OPC-only mix - be slowly leached from the cured
concrete. This decreased permeability is important
since it enhances resistance to chemical attack
(particularly by sulphates and chlorides).

The various cementitious reactions are all
exothermic but occur at different rates. During curing,
heat is generated within the concrete. In large volume
placements of OPC-only mixes, the more rapid reactions
can give rise to large temperature differentials
(40-70°C) resulting in cracking. PFA not only reduces
the quantities of OPC and hence heat released, it also
slows the damaging fast OPC—hydration reactions. This
also reflects in an initially lower compressive
strength, which however exceeds that of Portland cement
at later ages. There are clearly advantages from this
in making massive concrete structures, where internal
heat release might induce stresses.

The physical shape of PFA particles also plays a
significant part in its usefulness. Having been
produced in a furnace at 1400-1600°C the majority of the
particles are spherical. This property is valuable as
it allows the use of a lower water/solids ratio in the
mix without altering its handling properties. This has
the advantage of improved strength development, less
bleed and shrinkage and lower permeability, leading to
greater durability.

Minnick et al. (3) found a statistical correlation
between the water demand in concrete for a given
workability and the fineness and carbon content of the
PFA. Fineness was measured as % retained by weight on a
45 μm sieve. Unburned carbon tends to have a much
greater surface area than the ash with which it is
associated. Consequently surface area measurements can
be a misleading guide to fineness of material and sieve
analysis is considered more dependable. It is
fortuitous that fineness of the ash and low carbon
content are advantageous both to the chemistry and the
physics of using PFA as a partial replacement for
Portland cement.

Hardened Portland cement concrete can also suffer
from chemical attack by sulphate ions which can
(i) combine with free calcium hydroxide in the cement to
form gypsum and (ii) react with the calcium aluminates
to produce the mineral ettringite ($3\ CaO.Al_2O_3\ 3\ CaSO_4$
$32\ H_2O$) both of these reactions are accompanied by a
volume increase - with a resultant danger of disruption

of hardened concrete. Therefore partial substitution of Portland cement by PFA improves resistance against sulphate attack by reducing free calcium hydroxide in the system, which in turn reduces the solubility of calcium aluminates and the formation of ettringite (4). The reduced permeability, already mentioned above also contributes to added resistance to other chemical attack.

The use of PFA/OPC concretes in situations where freeze/thaw exposure is high has been avoided in the past because of its reduced resistance in these conditions. However recent research into the freeze/thaw resistance of both OPC and PFA/OPC concrete indicates that full protection against damage can be achieved by the use of air entraining agents producing much lower quantitites of finely dispersed air in the hardened concrete than previously specified.

In concrete, steel reinforcement corrosion (e.g. by chloride ions) is passivated at the high pH provided by the lime content of OPC. Concern has been expressed that the PFA reaction with the free calcium hydroxide might reduce the passivation ability of the concrete. However the results of recent research (6) into the long-term durability of PFA concrete shows clearly the benefits in protection to reinforcement by the inclusion of PFA in concrete mixes.

The alkali silica reaction (ASR) is another expansive degradation process in concrete. It is specific to a number of reactive aggregates and the alkali content of the mix plays an important role. The mechanism(s) are not understood fully but a high alkali content (usually from the cement) results in a high pore fluid pH which probably initiates a chemical attack on the aggregate. Whether or not the resultant reaction products are expansive is thought to be influenced by the local CaO/SiO_2 ratio. PFA in the mix appears to reduce both pH and CaO/SiO_2 ratio with the result that, as far as can be determined, no cases of ASR damage have occurred in PFA/OPC mix concretes in this country.

PFA as In-fill

Use of PFA as a civil-engineering in-fill accounts for around 20% of the total sales volume of the material used outside the ESI. Generally conditioned ash is used but additions of lagooned PFA can be made in order to adjust the moisture content. Because of the particle

shape the compacted material contains air voids of ∿10%
of the total volume and these, combined with the
comparatively low specific gravity (2.0 to 2.4), lead to
a low bulk density which is one of the properties making
the material such an attractive fill, especially over
poor grounds. Since the late 50's the material has been
used widely as embankment fill, fill behind bridge
abutments and other structures, particularly on
motorways. It can be compacted using vibratory rollers
or rubber tyred rollers. It has considerable
advantages, other than density, over clay because of its
resistance to settlement, largely due to its
self-hardening tendencies. These are caused by the
water soluble components calcium and sodium sulphate
which, with the optimum moisture content, form
cementitious bonds with the other components present in
the ash. Lagoon ash in which the water has not been
recirculated has a lowered concentration of these
solubles and so its hardening characteristics are
reduced.

The moisture content is a major factor in
determining optimum performance but it is not too
sensitive, provided the moisture falls within the range
0.8 to 1.2 times the optimum (determined by BS Test
1377: 1975 test 12). Another advantage of PFA in-fill
is, that with the addition of a thin layer of topsoil it
will support green growth.

PFA in Grouts

The pozzolanic properties, fineness and general
spherical shape of PFA particles also make it a
particularly suitable material for grouting and ∿10% of
PFA sold is used for this purpose.

In the broadest sense of the term grouting
PFA/water mix alone can be used for filling large
cavities but its more usual uses as grout are as
admixtures with lime, Portland cement etc. Lime
particles are finer than PFA so they tend to stabilize
slurries of the two which improves pumpability.
However, the rate at which its strength increases with
time is less than for PFA/cement mix. Portland cement
particle size distribution is similar to that of PFA so
the latter can be used as a filler in cement grouts
without further limiting the size of cracks/cavities to
which it can be applied. Indeed, as with concrete, the
presence of PFA improves the flow properties and hence
the ease of pumping and the depth of penetration. A

wide range of strengths can be obtained depending on the cement/PFA ratio so the mix can be accurately tailored to requirement. Where bleeding of cement grouts is a problem, addition of clays such as kaolinite or bentonite will make the grout more viscous for the short time required for the cement to develop an initial strength.

Lightweight Aggregate

Lightweight aggregate is a sintered rather than a cementitious product. PFA and coal dust are mixed and caused to agglomerate by controlled wetting and tumbling until the pellets of the required size are formed. These are then fired to give a carbon-free sintered product.

It is widely used for the manufacture of lightweight insulating blocks which also have the properties of high strength and inherent fire resistance. Large quantities are also used in structural concrete, floor screeds and roof screeds, where its strength and low density are particularly useful properties.

Building Blocks

Building blocks take most of the FBA output (2 Mte/yr) and over 1 Mte/yr of PFA. This represents the largest single use of power station combustion products. Dense aggregate blocks are made from FBA with additions by pressing the mix into moulds. Use of some PFA results in blocks with lower density and generally better thermal and mechanical properties; the latter can be controlled by changing the mix proportions. Where low density and high thermal insulation is required the inclusion of aluminium powder in the mix gives aerated blocks which are light and easily handled, have high insulating properties and are easily cut to shape for building into load bearing, non-load bearing and partition walls to meet modern Building Regulations. The aluminium powder in the mix reacts with alkaline pore fluid to produce hydrogen gas bubbles in the plastic mix before curing.

Brickmaking

Work in the fifties and sixties by the Building Research Station, the then British Electricity Authority and the British Ceramic Association demonstrated that

durable bricks could be made almost entirely from PFA.
However, these findings had little positive effect on
the brickmaking industry in that era of cheap energy and
a buoyant market.

Today, process energy is a significant proportion
of manufacturing costs and as PFA is a precalcined,
preground raw material its usefulness in brickmaking is
becoming accepted. Knowledge is now available on the
effects of carbon content and particle size to enable
commercial quality bricks to be made from a wide variety
of common brickmaking clays containing from 25 to 80%
PFA(7). The replacement of part of the clay with PFA
will also reduce brickwork emissions of SO_2 and
fluorine.

Cenospheres

Whilst cenospheres only form around 1% of the PFA
market their physical properties differ markedly from
the bulk of the ash and they certainly produce the most
interesting micrographs. Cenospheres are that portion
of the PFA with a density of less than 1.0. When PFA is
lagooned the cenospheres float and are skimmed off the
surface. They have the same chemical composition as the
bulk of the ash and differ only in that they are
thin-walled hollow spheres rather than solid or with
only a small central void with specific gravity in the
range 1.2 to 2.1.

Their combination of chemical and physical
inertness with low density makes them particularly
useful for lightweight insulation or fire resistant
structures. For example, in lightweight fire resistant
boards they are a superior alternative to pumice,
vermiculite, expanded clays, etc. in that they do not
compress and do not adsorb water. Graded cenospheres
produce good void filling because of their controlled
size.

They have also found many other applications in
moulding and in combination with synthetic materials to
produce stronger, lighter structures.

Mineral Extraction

As Table 1 shows, PFA contains high concentrations
of some useful elements and much research work has been
done on their extraction, particularly alumina. The
limestone PFA sinter method involves the high

Table 1   Constituents of PFA

Constituent	Range (%)	
Silica SiO_2	47	- 52
Alumina Al_2O_3	27	- 32
Iron Oxide Fe_2O_3	7	- 14
Calcium CaO	1.5 -	5.0
Magnesium MgO	1.0 -	2.0
Sodium Na_2O	0.5 -	1.5
Potassium K_2O	2.0 -	4.0
Titanium TiO_2	0.8 -	1.2
Total Sulphate SO_3	0.2 -	1.2
Water Soluble SO_3 (2:1)	0.2 -	2.8 g/l
Loss-on-Ignition	2	- 12%
pH	6	- 11.5
Specific Gravity	2	- 2.4

Table 2   Estimated Input and Outputs (per an.)
for Drax PS

(4000 MW(e), 1.7% S coal, 82% load factor)

IMPORTS	EXPORTS	
Limestone 480 Kte	Gypsum	860 Kte
	Gypsum Sludge (from waste water treatment)	10-50 Kte

temperature (1100-1300°C) sintering of these two
materials with the subsequent production of alumina.
American workers (8) have recently shown that the
limestone can be partially replaced using sludge from
FGD plant. A more recent approach has used alkali
pressure leaching of the ash. As yet, however, the
comparatively stable price of world market alumina and
the high energy and capital costs of new plant have
prevented significant commercial application. Some of
the minor elements, present, such as gallium and
vanadium also have potential value but again the
commercial scene is not yet ripe for their further
development (9). Their future potential is not being
ignored, however and work continues to be funded at
universities in the area of aluminium recovery,
utilizing FGD pre-scrubber acid and other waste acids.

Disposal

     The disposal of the ash product is a major
consideration in power station operation and the
situations can vary from that where a local manufacturer
of PFA products can take all the ash the station
produces to that where the bulk of the ash must be sent
to landfill. Even in the former case it is unwise to
assume the manufacturer can always accept total
production and contingency plans must be in hand.

     A typical case is the disposal of ash from
Ferrybridge and Eggborough power stations (10) which
between them may produce around 2 Mte of ash per annum.
No suitable quarries, clay or sandpits were available in
the vicinity so planning permission was obtained for the
building of a landscaped mound on Gale Common, an area
of land which is liable to be affected by colliery
subsidence. The ash is transported to the site as a
slurry by pipeline. It is initially lagooned before
being used in mound construction.

     An ash mound is being constructed at Drax Power
Station but here the ash is conditioned with 12% water
and carried by belt conveyer to the site at Barlow where
it is disposed as a solid.

     3 THE IMPACT OF GASEOUS EMISSION CONTROL ON SOLID
       COMBUSTION PRODUCTS

The control of gaseous pollutants, particularly of
sulphur dioxide, from fossil-fuel burning power plants
was pioneered in the UK in the 30s, when once-through

spray washing towers using river water were installed
below the chimneys of Battersea power station and at
Bankside PS. After the Second World War power station
building moved out of the towns and closer to the
coalfields. In the early sixties the size of planned
stations (2000 MW and even 4000 MW) increased interest
in possible control measures. Research work was carried
out during this period on dry (adsorption) methods and
in parallel the development of tall stacks with wide
dispersion of pollutants was pursued. This latter
technique proved most successful at reducing ground
level concentrations in this country and was widely
adopted here and elsewhere. In the early seventies
evidence emerged of the link between fossil-fuel burning
and acid rain, to which trans-frontier movement of acid
gases was a significant contributor. Development of
large scale flue gas desulphurisation methods followed
in Japan, Western Europe and the USA and there are now
around 100,000 MW(e) of plant fitted with FGD throughout
the world.

A major requirement which has emerged during this
development has been that processes which reduce gaseous
emissions should not produce by-products which present a
disposal problem. Thus the most favoured methods for
SO_2 control are those which convert flue gas sulphur
dioxide into a marketable product e.g. elemental
sulphur, sulphuric acid or gypsum, the relative
saleabilities depending on the location of the
individual power plant.

The UK, in concert with all the EEC countries has
undertaken to reduce its SO_2 emissions, with respect to
1980 figures, by 60% in 2003 with intermediate stages of
20% and 40% in 1993 and 1998 respectively. A major step
is being made by National Power towards these goals by
the construction of FGD plant which has commenced on the
4000 MW plant at Drax, Yorks. The limestone/gypsum
process has been selected for this installation on the
basis that it is the most well developed and reliable of
available technologies worldwide.

As is shown in Table 2, for Drax this results in
the consumption of ∿500 Kte of limestone and the
production of ∿800 Kte of gypsum per annum. This
material is to meet commercial specifications in terms
of purity mainly for wall board and perhaps some bag
plaster, for which purposes it will be sold. However,
the gypsum production from this one station, albeit the
largest in Europe, is expected to account for ∿30% of

Table 3   Barlow PFA/Gypsum Experimental Mounds

Mound	Ash/Gypsum Ratio	Lime Content
1	2:1	0%
2	2:1	1%
3	1:1	0%
4	1:1	2%
5	100% Ash	-
6	100% Gypsum	

current UK gypsum utilization and perhaps 25% at the
time the plant is fully commissioned and so other
disposal routes may be necessary for excess or
out-of-specification material.

This problem is under consideration and three major
routes are being examined; codisposal with unsold PFA;
agricultural use in the conditioning of poor soils and
the reclamation of saline soils; the production of
gypsum/PFA/cement blocks for application in marine reef
building.

Co-disposal

The use of PFA for landfill in the reclamation of
old quarries and brickworks claypits is widespread, as
is landscaping (see 2.9 above).

A large-scale investigation into the possible
effects on the environment of gypsum/PFA co-disposal is
proceeding at Barlow, adjacent to Drax Power Station,
where a number of experimental mounds has been
established to investigate the chemical and physical
stability of different ash/gypsum mixtures. Six mounds
of varying gypsum, ash and lime content have been
established (Table 3). The lime additions were to the
mixtures lower in ash where the alkaline content (for
bonding) might otherwise be too low. A comprehensive
programme of analysis of rain water run-off and leachate
passing through the mounds is under way to determine the
levels of heavy metal transport and shear tests are
being carried out to determine stability.

Agricultural Use

(i) Non-saline soils

The use of gypsum on non-calcareous soils in the UK improves the physical properties of the soil, that is, better workability, permeability, stability and seed bed quality. In the long term this is likely to increase the yields of the fields so treated.

(ii) Saline soils

Gypsum also improves the physical properties of saline, or salt affected soils with a resultant increase in crop yield. There are around 2000 km^2 of salt-affected soils in England and Wales and if only a tenth of this is treated (at 20 te per hectare once every five years) a demand well in excess of 1 Mte per year would result. At present, phosphogypsum, a waste product from phosphate fertilizer production is used but cannot be spread at greater than 10 te per hectare because of its impurity content. The high purity of FGD gypsum will cause no restriction.

Cement Stabilized Blocks for Use in Artificial Reefs

EPRI (11) have carried out extensive trials on the evaluation of stabilized FGD materials in the construction of artificial reefs and this option is also being followed vigorously in this country. A variety of gypsum/ash mixtures with between 5 and 10% added cement content have been formed into blocks and their characteristics examined. The mixtures showing the best strengths, combined with resistance to sea-water exposure have been used to prepare an artificial reef composed of six piles of approximately 8 te each, which was set up in Poole Harbour in mid-1989. All the indications to date are encouraging; the reefs were quickly colonised by marine life and have also shown stability under the very rough weather conditions experienced at the start of 1990.

Work on the chemistry, physics and ecological impact of the reefs will continue for some time.

Waste Sludge

In early US limestone FGD processes, the oxidation stage following SO_2 absorption was not complete resulting in a thixotropic, mainly calcium sulphite product which could only be dumped. The adoption of a

gypsum producing process largely removes this problem
although a relatively small quantity of sludge - mainly
gypsum of small crystal size and calcium carbonate is
produced from the waste water treatment plant of the FGD
process. One obvious method of disposing of this sludge
is by co-disposal with excess fly-ash, as described
above for gypsum. For similar reasons the effects of
leaching on PFA/sludge mixtures are being investigated.
A second possibility already in use at some power
stations on the continent is sludge refiring. The
sludge can either be dispersed on the coal before it
enters the combustion train or actually be injected into
the furnace volume at a controlled rate so that it
becomes incorporated into the ash with minimal effect on
the ash properties.

     Both routes are being actively considered at the
present time. There is a possible disadvantage from the
presence of fluorides in the coal. Hydrogen fluoride in
the flue gas is removed in the FGD plant as calcium
fluoride which is then precipitated in the water
treatment plant, ending up in the sludge. Refiring may
result in the reformation of hydrogen fluoride and the
consequent recycling would build up the calcium fluoride
in the system. If this is the case, a calcium
fluoride-rich removal stream would have to be set up in
the water treatment plant.

     4 DE-NO$_x$ PROCESSES

Processes for reducing the level of nitrogen oxides
emitted from power stations fall into two categories;
combustion modifications and chemical reduction methods,
both of which can have some effects on the PFA product.

Combustion Modifications

     Nitrogen oxides are mainly formed from
(i) oxidation of the organo-nitrogen compounds in the
coal (fuel-NO$_x$) (ii) the $N_2 + O_2$ reaction at flame
temperatures (thermal NO$_x$). The former can be much
reduced by 'staging' the combustion. Low NO$_x$ burners
allow the fuel to first burn in an oxygen lean
environment when the fuel nitrogen compounds, which are
volatile, produce radicals which interact with each
other to produce nitrogen rather than oxidise. The
second stage involves an oxygen rich environment which
burns the fuel completely without NO$_x$ production.
However there are examples where the installation of
these burners results in a slightly less efficient

burn-out of the fuel with a consequent increase in the
level of carbon in the PFA. This of course has
repercussions for the use of the PFA where (other than
for brickmaking) the presence of carbon may be
deleterious.

Chemical Methods

These involve the reduction of nitrogen oxides with
ammonia or amine containing compounds to produce
nitrogen and water.

$$4NO + 4NH_3 + O_2 \rightarrow 4N_2 + 6H_2O$$

The reaction can occur at high temperature (900-1100°C)
without catalysts or at lower temperatures (350-400°C)
utilizing a catalyst. In either case the possibility of
ammonia slip exists which could produce ammonium
bisulphate or sulphate by reacting with sulphur oxides
in the gas phase, some of which could end up in the PFA,
or direct adsorption of ammonia onto the PFA surface
could occur. Either may affect the properties of the
ash. German sources indicate that a proportion greater
than 5 vpm slip in the gas phase can cause problems with
the subsequent ash handling, because of ammonia release
on water addition.

5 SUMMARY

Whilst many millions of tonnes of pulverised fuel ash
are being utilized, mainly in the construction industry,
this still only accounts for ~50% of that produced and
there is clearly room for further utilization.

The advent of flue gas cleaning processes has added
a further dimension to the problem of solid disposal
from power stations but this is migitated by producing
marketable products or disposal methods which give
environmental improvements.

ACKNOWLEDGEMENTS

The authors are grateful to their colleagues in National
Ash for their helpful comments.

This paper is published with the permission of
National Power PLC.

REFERENCES

1. H.S. Meissner, 'Pozzolans used in Mass Concrete',
    Symposium on use of Pozzolanic Materials in Mortars
    and Concretes, Special Technical Publication, 1949,
    No. 99, ASTM, 16-30.
2. FGD Handbook, IEA Coal Research ICEAS/BS, London
    1987.
3. L.J. Minnick, W.C. Webster and E.J. Purdy,
    'Prediction of the Effect of Fly Ash in Portland
    Cement and Concrete, J. Mats, 1971, 6(1) 163.
4. G.L. Kalonsek, L.C. Porter, E.J. Benton, 'Concrete
    for Long Term Service in Sulphate Environments',
    Cement Concr. Research, 1972, 2, 79.
5. Cement Research of India, 'Incidence of Corrosion
    of Steel Reinforcement in Fly Ash Concrete, 1974,
    Report No. RB-3-74.
6. Z. Scislewski, 'Effect of Fly Ash on the Protection
    of Concrete Reinforcement', Cahiers du Centre
    Scientifique et Technique du Batiment, 1975, No.
    165, Cahier, 1351.
7. M. Anderson and G. Jackson, 'The History of PFA in
    Brickmaking in Britain', Trans and J. Inst.
    Ceramics, 1987, 8 (4), 99.   s
8. G. Burnet, M.J. Murtha and N. Harnby,
    'Co-utilization of Pulverised Coal Ash and Flue Ga
    Scrubber Sludge', 2nd Int. Conf. on Ash Technology
    and Marketing, London, 1984.
9. J.M. Noone, 'Mineral Extraction from Fly-ash',   l
    ibid.   .
10. J. Brown and P.M. Owens, 'General and Environmenta
    Aspects of Ash Disposal Allied to Land Reclamation
    2nd Int. Conf. on Ash Technology and Marketing,   ,
    London, 1984.
11. EPRI, 'Coal Waste Artificial Reef Programme', 1985
    Report No. CS 3936.

The Environmental Effects of Oil and Gas Production

D. E. Martin

BP INTERNATIONAL PLC, BRITANNIC HOUSE, MOOR LANE, LONDON EC2 9BU, UK

INTRODUCTION

Crude oil has been produced commercially in the UK since World War 2, from small fields mainly in the East Midlands. In the 1960s, attention shifted to the North Sea, first to gas fields in the southern sector, and then in the 1970s to the oilfields of the northern sector. Development of these resources was rapid so that by 1986, production had risen to about 177 million tonnes of oil and 96 thousand million m^3 of natural gas. Together these provided the same energy as about 400 million tonnes of coal, approximately the quantity then mined throughout Western Europe. Overall, the North Sea oil and gas industry provides about 20% of Western Europe's energy requirements.

UK onshore oil production is much smaller at less than 1,000,000 tonnes per year. This will increase this year to a rate of over 3,000,000 tonnes per year when the Wytch Farm oilfield in Dorset comes on stream.

The technology for producing oil offshore and onshore is similar. However the different locations can lead to differences in the environmental impacts, and the measures taken to reduce them. I will deal firstly with offshore developments and then consider the special problems which arise in the development of a major oilfield in a sensitive area such as Dorset. Firstly, I would like to consider environmental policy.

ENVIRONMENTAL POLICY

The oil industry sees concern for the environment as being of fundamental importance. To use as an example, my own company BP has a policy that states that it "Will endeavour to limit any adverse effects in the physical environment in which its activities are carried out". This was summed up by one of our Managing Directors in the statement "Environmental protection is an integral part of running the business".

BP's commitment to environmental protection stems from a number of reasons: legal requirements, public relations, good relations with authorities, cost saving, and not least because many BP employees, shareholders and customers have a genuine concern for the environment. One means to fulfilling the policy and ensuring that adequate measures to protect the environment are applied is by using the techniques of Environmental Impact Analysis or EIA.

In a typical oilfield development, this process is applied at a number of stages: selection of areas to apply for licenses, selection and design of drill sites, preparation of oil spill contingency plans, design of production facilities, and on land, when submitting planning applications. Once the field is in operation, monitoring and audits are carried out, and finally environmental issues will need to be considered for abandonment and restoration. BP has adopted the name "Environmental Protection Management" to cover the whole process. The EIA reports themselves can be considered a summary of the EPM process. They serve a useful function in explaining the environmental issues, both within BP and to interested external audiences. As they are designed to be read by non-technical persons, they are also often useful for describing the project to non-specialists within, as well as outside BP.

This procedure has proved to be of great value in ensuring environmental protection in a cost effective manner. In particular:

(i)   to avoid delay in the authorisation of projects caused by unforeseen environmental objections.

(ii) to insure against future liabilities resulting from environmental impairment.

Although there are many similarities between offshore and onshore developments, there are also differences. These two topics will be considered separately.

OFFSHORE DEVELOPMENTS

Exploration

The development of an oil or gas field passes through a number of phases, and the impacts observed are different in each phase. In the initial exploration stage, seismic surveys cause little effect now that low energy sound sources are used so that explosive charges are rarely required. Once a prospective structure has been identified, a well will have to be drilled, as this is the only way to prove the presence of oil or gas.

Exploration wells are generally drilled from mobile rigs mounted on dynamically positioned drill ships, anchored semi-submersible vessels or jack-ups depending on the water depth. Assuming that oil is discovered, further appraisal wells will be drilled by similar methods to determine the extent of the field. At this stage, the impacts from drilling are limited to a small area around each wellsite. The impacts from drilling will be discussed later.

Construction

When the presence of a field has been confirmed, the design of the permanent facilities can proceed. The "traditional" procedure is to construct a platform to carry both a drilling rig and production facilities, drill the wells required and produce the oil. There are now many variations on this procedure such as pre-drilling some or all the wells before installing the platform, floating production systems, sub-sea wells, sea-bed installations, etc. The choice depends on a number of factors including the size of the field, gas/oil ratio, depth of water etc. With most systems, a considerable construction effort will be required using a variety of vessels and the employment of several hundred personnel. Inevitably there will be some discharges to the sea, but in general these will have very minor effects.

In many cases, pipelines will have to be laid either to shore or to another field already served by a

pipeline. Pipelines are laid from a barge, and the
actual laying causes little disturbance. Sometimes the
pipeline will be trenched into the sea-bed, and digging
of the trench disturbs a large area of sea-bed, and
introduces sediment into the water column. The sea-bed
soon recovers and is rapidly recolonised from the
surrounding area.

The final stage of pipelaying involves pressure
testing with water to which low concentrations of
anti-corrosion chemicals have been added. This water
has to be discharged at sea, but the discharge is
controlled so that the chemicals are rapidly dispersed
to harmless levels.

Often the most difficult operation is bringing the
pipeline ashore without damaging the coastal margin.
This is particularly important in a country like the
Netherlands where the coastal sand dunes are a vital
element in protecting the country from the sea. One
method of overcoming this problem was used recently by
BP for a gas field off the Netherlands. In this case,
the pipe was pulled through a hole drilled horizontally
right under the sand dunes and the beach, so that the
dunes were not disturbed at all.

One problem arising from construction and
pipelaying is that pieces of equipment get dropped or
fall overboard. If left on the sea-bed, they can damage
fishing nets. It is therefore standard practice at the
end of the construction period to survey the sea-bed
around platforms and along pipeline routes and recover
any debris found. Even so, nets do get damaged. Where
the company responsible cannot be identified, national
schemes are funded by the oil companies to compensate
fishermen for loss of gear or catches.

Where not trenched, the pipelines will remain on
the sea-bed throughout the life of the field. So far,
over 9,000 km. of pipelines have been laid. These can
be a hazard to fishermen for many years. Their
positions are of course marked on charts but surveys
often reveal nets tangled around pipelines. This may be
because fish are attracted to pipelines and hence make
the routes a profitable place to fish. Needless to say,
the pipelines are designed to be strong enough to resist
damage from fishing gear.

A further impact on fisheries arises from the
presence of the platforms themselves. For safety

reasons, a 500 m. exclusion zone is imposed, within
which ships, apart from those associated with the
platform, are banned. For the 80 or so oil platforms in
the North Sea, this amounts to some 60 km^2 where fishing
is banned. However, this only represents about 0.01% of
the North Sea. Probably a similar area is affected by
gas platforms.

Drilling

Drilling of the production wells is probably the
source of the greatest impact on the environment. From
an average well, about 500 to 1000 m^3 of rock cuttings
will be discharged, and a typical platform may require
30 wells. The only practical disposal route for this
material is to discharge it onto the sea-bed under the
platform. All told, some 5,000 wells have been drilled
to date.

During drilling, a working fluid known as 'mud' is
circulated through the hole. Despite its prosaic name,
drilling mud is actually a complex mixture carefully
formulated to fulfil a number of functions. These
include ensuring the safety of the well by stopping
fluids entering from the formation, lubricating the bit,
and removing the cuttings.

The major solid components are natural clays (e.g.
bentonite) and barytes. The liquid phase is either
composed of sea water (water based muds) or an emulsion
of water in oil (oil based muds). Which ever type is
used, the discharged cuttings will be coated with mud.

There are a number of reasons for using oil based
muds, but one of the principle reasons in the North Sea
is concerned with the chemistry of the rocks which have
to be drilled through.

Under the North Sea there are long sequences of
clays, shales and mudstones. When exposed to water, they
become hydrated and swell. This can lead to problems of
clogging in the equipment used to separate the mud from
the cuttings and return it to the well.

The main difficulty however is that the rock lining
the well bore also swells, so that the bore narrows. In
extreme cases, the bore can narrow so much that the
drill pipe becomes stuck and can shear off. Apart from
being inconvenient, if this happens when the drill is
passing through a high pressure stratum, the risk of a

blowout is increased while fishing operations are
undertaken to recover the bit and drill string. If this
fails, the bottom part of the well will have to be
plugged, and the well sidetracked around it.

An alternative to oil based muds is to limit
hydration by increasing the ionic strength of the mud
fluid with chemicals such as potassium chloride. This
is less effective than oil, but is often adequate to
allow vertical or slightly deviated drilling. However
it is not suitable for drilling the long highly deviated
wells necessary to reach the edges of a reservoir from a
platform. It would be feasible to drill all wells
vertically from locations all over the field. However,
this would mean drilling, wellheads and pipelines all
over the field, affecting a wide area of seabed, and
increasing the chances of leakage or a major spill.

The main environmental difference between the mud
types is the rate of recovery after drilling ceases.
With water based muds, the comparatively inert cuttings
will soon be mixed into the sea-bed and colonised by a
normal benthic community. With oil based muds, recovery
will take much longer because of the slow biodegradation
of the oil on the cuttings. This oil is tightly bound,
and little is desorbed into the sea and slicks are not
normally observed. The main effect arises from the
organic enrichment of the seabed.

A number of measures are taken to reduce the
impacts arising from the use of oil based mud, and
research is under way on methods to reduce the quantity
of oil discharged. One development made a few years ago
was to change the oil phase from diesel to less toxic
highly refined mineral oils. In these, the aromatic
hydrocarbons are selectively removed, and it is these
aromatics which are the most toxic to marine life.

An additional tactic which is being tried is to
reduce the oil content of the liquid phase. Muds with
less than 50% oil have been used, but these are
formulated to ensure that oil remains the continuous
phase so that water cannot reach the rock structure.

On many drilling rigs, the cuttings are washed
before discharge using various solvents or surfactant
solutions followed by centrifuging. Although some oil
can be removed and recovered by these systems, none is
very effective, and research has been progressing for a
number of years on novel cleaning agents. The problem

is to find a medium which is both effective and less toxic than oil. As the toxicity of oil is low, and it exerts most of its effects by acting as an oxygen scavenger, this is quite a challenge.

Other solutions suggested have been incineration or distilling off the oil, but neither has yet been proven in the conditions experienced offshore. There is also the possibility of shipping the cuttings to shore for treatment or disposal, but apart from the difficulties and dangers attached to this procedure, it is really only shifting the problem elsewhere. A simple tactic used by BP in some of its recent fields is to reduce the diameter of the wells. This can halve the volume of oil discharged.

Operations

During the production stage, the impacts become much less, particularly in relation to the volume of oil or gas being handled. The flares on oil platforms may look spectacular, but on large fields most of the gas is either used as fuel on the platform or piped to shore, and only the minimum amount required for safety is flared. On smaller fields, or those with a low gas/oil ratio, excess gas is flared as it is uneconomic in energy as well as money terms to lay a pipeline to bring the gas ashore.

Most North Sea gas and oil is low in sulphur. There are however a few fields where there are significant concentrations of H_2S and CO_2 in the gas. These are removed by counter current extraction with an amine such as diethanolamine. The H_2S is stripped from the solvent and oxidised in a limited supply of oxygen so that about one third is converted to SO_2. The H_2S and SO_2 then combine in the Claus reaction

$$2H_2S + SO_2 \text{ ----> } 3S + 2H_2O$$

producing elemental sulphur which is usually sold for sulphuric acid production.

The gas must also be dried before it is piped to shore by extracting the water with a glycol such as triethylene glycol. If this were not done, at sea-bed temperature and at high pressures, solid methane hydrates would form which would plug the pipeline.

As with most other industries, chlorofluorocarbons and halons are used in air conditioning plants and for fire-fighting. With the current concerns over the discharge of these materials, BP is looking at ways to minimise the use and losses of these materials.

Various aqueous discharges are made to the sea. Cooling water will be at an elevated temperature and contain small quantities of chlorine, but is otherwise sea-water. Sewage and domestic waste water are generally discharged untreated apart from maceration, but the amount from the 150 or so people on a platform is negligible compared to the volume of the sea. Platform drains are designed to minimise the amount of oil entering them. Before discharge, the effluent is further treated to remove oil.

In most oilfields, although dry oil is produced initially, after a time, water is produced as well. As the field ages, the proportion of water increases. This water is separated from the oil and discharged to the sea after treatment. Because of the limited space and weight restrictions on a platform, it is not possible to treat this water to the same standards as might be expected onshore. The most common types of equipment used are tilted plate interceptors and induced gas flotation, but hydrocyclones are now beginning to be used. This equipment is usually able to reduce the oil content to below the 40 mg/l oil in water standard imposed by all the North Sea states.

Although we all know what oil is, to define it is actually quite difficult. For practical purposes, it can only be defined by an analytical method against a specified reference oil. The method specified in the North Sea is by infra-red absorption of a Freon extract at 3.42 microns which detects CH bonds. In some fields, under the conditions of test, organic acids are also extracted and are measured as oil even though they are water soluble. This can be overcome by extracting the acids with activated silica.

Although physical separation methods are used to remove oil, the separation is as much a chemical process as physical. Invariably, chemical additives have to be used to flocculate the oil present as a dispersion of micron sized oil droplets stabilised by surfactants naturally present in the oil. Polyelectrolytes are used for this purpose together with foaming agents in induced gas flotation equipment.

Scale is also a problem in production equipment.
North Sea formation water is typically high in Barium
and Strontium. When this comes into contact with sea
water containing sulphate, precipitation of insoluble
sulphates occurs. Scale formation is also caused by
changes in the HCO_3^-/CO_3^- equilibrium. As the pressure
is reduced, CO_2 comes out of solution and calcium
carbonate is precipitated. To reduce this, various
scale inhibitors are used both in sea water injection
wells and in production equipment.

Measures to prevent scale formation are taken not
only to reduce plugging of equipment. The reservoir
rocks contain small quantities of uranium and thorium
minerals which results in very low concentrations of
radium in the formation water. This co-precipitates
with Barium and Strontium, and hence is concentrated in
the scale. The scale is therefore a low specific
activity (LSA) source with activities below 15,000 Bq/g.
Whilst still in the equipment this is quite safe as the
shielding provided by the pipework is more than adequate
to contain the radiation emitted. Special procedures
have to be adopted for vessel cleaning and entry to
minimise radiation exposure to workers. Disposal is
also a problem, the usual method being to grind up the
scale and discharge to the sea where it becomes widely
dispersed. Surveys around platforms have failed to
detect any significant radioactivity from this source.

Comparatively little waste material is generated on
platforms. Such wastes as are generated are returned to
shore for disposal by normal methods, except for food
wastes which may be ground up and discharged to the sea.
These will provide food for the marine life which
thrives on and around the platforms.

Oil Spillage

Oil spills into the sea are undesirable for obvious
reasons. In the short term, oil floating on the sea
poses a particular hazard to sea-birds. It can also
cause severe damage if it reaches the shore.
Fortunately, most of the North Sea oilfields are far
enough from shore to make this unlikely. In the long
term however, crude oil, being a natural product, will
be degraded by the natural bacteria in the sea.

With the volume of oil being handled, it is
inevitable that some will be spilt, despite all the
precautions that are taken. Most of these spills will

be small, less than a few tonnes, and will soon
disappear. There are occasional major accidents and
much research has been devoted to producing equipment to
recover oil from such incidents. BP maintains a large
stock of such equipment at the Oil Spill Service Centre
in Southampton on behalf of a number of oil companies.
Even so, in the event of a major spill in the open sea,
there is no way that all the oil can be recovered.

Good management has ensured that there have been
very few major spills in the North Sea, and these have
not had the serious effects observed from shipping
accidents such as the Torrey Canyon or Amoco Cadiz.
Certainly, studies of sea-birds have not shown any
evidence that their numbers have declined as a result of
oil activities.

The oil industry's Exploration & Production Forum
attempts to calculate the amount of oil lost to the
North Sea from the oil industry. The figures for 1985
which were presented to the 1987 London Conference on
the Protection of the North Sea Environment
were as follows:

Oil Spills	729 tonnes
Oil in Produced Water	2,122 tonnes
Oil in Ballast Water	98 tonnes
Oil in Terminal Effluent	296 tonnes
Oil on Drill Cuttings	25,594 tonnes
TOTAL	28,839 tonnes

The Institute of Offshore Engineering for the same
year estimated total oil inputs to the North Sea as
follows:

Natural Seeps	300 - 800 tonnes
Atmospheric Rain Out	19,000 tonnes
Run-off from Land	40 to 80,000 tonnes
Coastal Sewage Discharge	3 to 16,000 tonnes
Oil Terminals	6,000 tonnes
Industrial Effluents	9,000 tonnes
Accidental Shipping Losses	5 to 12,000 tonnes
Operational Shipping	No agreed figure
Offshore Production	23,000 tonnes
TOTAL	107 - 165,000 tonnes

These figures show that the oil resulting from
production operations is a fairly small proportion of

the total entering the North Sea. It should be stated that all these figures are estimates and may be open to error. However, perhaps surprisingly, analysis of North Sea water shows that oil levels are low, in fact lower than some measured in apparently pristine areas of the world's oceans.

Abandonment

At some stage, the oil and gas will run out, and the fields will be abandoned. There is currently a lively debate on what should be done with the platforms. Either total removal or toppling onto the sea-bed, as was done with Piper Alpha, are possibilities. In either case a sufficient depth of water for safe navigation will have to be left. Platform remains on the sea-bed will obviously be a hazard to fishermen's nets. On the other hand, the remains will form an artificial reef and may help to conserve fish stocks.

Whatever the ultimate fate of the platforms, before abandonment the wells will be plugged, and all oil, chemicals and hazardous materials will be removed so that the remains will not give rise to the threat of pollution.

Monitoring

It is impossible to draw any inference on the effects of oil production from general studies of the health of the North Sea as so many other factors and influences are involved. However, throughout the development of the North Sea oil industry, the oil companies have carried out extensive biological and chemical monitoring around their installations. They have also sponsored research on a wide variety of environmental matters concerned with the North Sea such as the distribution of sea-birds. These have greatly added to our knowledge of their life patterns and migration routes.

The monitoring studies have shown that where oil based muds have not been used, very little difference can be seen between the ecology of the immediate environs of the platform, and similar control areas some distance away. In fact, the platforms are usually host to a very healthy growth of mussels and other marine species.

Where oil based muds have been used, there is

undoubtedly damage to a small area of sea-bed. To put
this in perspective. A recent Environmental Impact
Assessment for a BP 70,000 bbl/day North Sea Oilfield
has predicted from analogy with other similar fields
that up to 20 ha. of sea-bed will be severely impacted,
but by no means sterile. This can be compared with the
Wytch Farm oilfield onshore in England where extreme
measures have been taken to protect the environment,
particularly by limiting land take. Even so, some 40
ha. of land have been effectively sterilised under
concrete for many years.

ONSHORE DEVELOPMENTS

As stated earlier, there are many similarities
between onshore and offshore developments. Basically,
the same technology is used for drilling and production,
although fortunately the geology onshore does not
usually necessitate the use of oil based muds. The main
differences arise from the different habitats
encountered and the presence of people onshore.

The North Sea is a fairly homogeneous environment,
and over large areas, there is little to choose between
one site and another. On land this is not true and even
neighbouring plots of land can have very different
environmental sensitivities. This gives the opportunity
to select areas of land to use where little
environmental damage will result. Even in the most
sensitive areas, there are usually patches of lesser
value.

Wytch Farm Oilfield

An example of developing an onshore oilfield is
given by Wytch Farm in Dorset. This field was
discovered by British Gas (with BP as partners) in 1974
near Corfe Castle in Dorset. The field was initially
developed to produce 4000 barrels a day of oil, which
although small by North Sea standards made it the
largest onshore field in the UK. Later, a much bigger
reservoir was discovered at a greater depth, which it
was believed extended under Poole Harbour. However,
little development was carried out as the Government was
in the process of requiring British Gas to sell their
share. Eventually in 1984, a consortium of small
companies bought out the British Gas share and BP became
operator.

BP were faced with three main problems:

(i)   To determine the actual extent of the reservoir.

(ii)   To design and build facilities to extract and process 60,000 barrels a day of oil plus LPG and Gas (a major field by any standards).

(iii)   To transport the oil to a refinery or shipping terminal.

These had to be carried out in a designated Area of Outstanding Natural Beauty and Heritage Coast, which is liberally sprinkled with National Nature Reserves and Sites of Special Scientific Interest (SSSI) administered by the Nature Conservancy Council, various other protected areas such as RSPB and local naturalist trust reserves, National Trust land and scheduled ancient monuments. It is also a tourist area popular for its 'unspoilt character'. Much of the area is lowland heath, a habitat which supports many rare species and is fast disappearing under roads and housing in Southern England.

One of the main concerns of the planning authority (Dorset County Council) was that the development should be inconspicuous or preferably invisible. A difficult operation when the whole area is overlooked by Corfe Castle, one of the most popular tourist attractions in southern England. Fortunately (for BP), after the war, large areas of heathland had been covered in forestry plantations, and BP were able to use these to screen most of the installations. This screening has to be maintained for many years, so management agreements have been made with neighbouring landowners to ensure that the tree screens are kept beyond the age at which they would normally be clear felled.

The phasing of the development presented BP with an opportunity. The first phase to appraise the reservoir involved drilling on a small island in Poole harbour called Furzey Island. BP produced an EIA document, much in its usual style and was somewhat surprised when this was heavily criticised on the basis that it was inadequate and contained no predictions. In fact when we went through it, we counted about 80 firm predictions, but many of these were predictions of minimal impact which were not believed.

When planning permission was applied for, many of

the statutory consultees objected, but in nearly all
cases, the objections were on the grounds of principle.
However permission was granted and drilling went ahead.
It was now that the wisdom of the phased approach became
apparent as BP was able to demonstrate that it could do
the job properly, and that the predictions of minimal
impact made in the EIA were correct. In particular that
a drilling rig, although apparent, was not obtrusive,
that the noise of drilling could be controlled and did
not disturb either birds or tourists; and most
importantly no oil was spilt into the harbour.

      While the work on Furzey Island was proceeding,
design work on the main development plan was proceeding.
Close contact is always maintained in BP between the
engineering design staff and the environmental advisers.
Environmental Assessment is seen not just as a way of
assessing the effects, but as an integral part of the
design effort where changes are continually made for
environmental reasons with the aim of minimising the
impact. For Wytch Farm, this procedure was taken even
further than usual with a BP environmental specialist
seconded full time to the design team at the
contractor's office.

      As a result of this a large number of changes were
made to the design for environmental reasons including
two major changes involving moving the LPG storage from
one installation to another, and a complete redesign of
the largest installation at a very late stage to reduce
its visual impact. Interestingly, the component of the
development which caused more heart-ache and effort than
any other was the provision of access to the site and
the design of a road.

      In fact there were very few instances of major
innovation necessary to develop the field. The measures
normally taken in good oilfield practice to minimise
losses of valuable oil and to ensure safe operation also
reduce environmental impact. The extra dimension that
was needed was careful planning of both equipment and
techniques to ensure that the operation would blend
sympathetically into the environment.

      This is not to say there were not many problems to
solve. For example; (i) how to get the oil from Furzey
Island to the mainland which was solved by the
horizontal drilling technique mentioned earlier; (ii)
how to dispose of all the drill cuttings, not oily but
in large quantity and in the form of a wet slurry.

Suitable landfill sites were found; (iii) how to dispose of produced water, solved by re-injecting it into the reservoir to help push out the oil. Unfortunately, this would not supply enough water so sea water will also be required. The problem of scaling was overcome by dividing the reservoir so that formation water will be pumped into one half and sea water into another.

The biggest remaining problem was how to get the oil to market. The logical solution was a pipeline to the Esso Refinery at Fawley and an existing BP Shipping terminal at Hamble on the other side of Southampton Water. This meant crossing the Perambulation of the New Forest although it avoided the SSSI heartland areas. BP looked at other pipeline routes, and other interested parties (Poole Harbour and British Rail) put forward their own schemes. Dorset conducted a public consultation process which finally opted for the New Forest Route, and detailed studies produced a route with minimum impact which was approved following a public enquiry.

At the time of writing, the new facilities are about to come on stream and our predictions of minimal impact will finally be put to the test. That is not the end of the story however. Since designing the first stages, BP has now discovered that the Wytch Farm reservoir extends well out into the sea across Poole Bay. There is now the problem of developing an offshore field, not 100 miles out in the middle of the North Sea but one or two miles off Bournemouth Beach. This will pose all sorts of environmental problems which we are currently investigating.

Vehicle Emission Control Technology

J. M. Dunne
DEPARTMENT OF TRADE AND INDUSTRY, WARREN SPRING LABORATORY,
GUNNELS WOOD ROAD, STEVENAGE, HERTFORDSHIRE SG1 2BX, UK

1 INTRODUCTION

Responding to growing pressure within the European
Community for a more rapid pace to emission reduction
controls from motor vehicles, a series of Directives
have been issued since 1988 mandating much more severe
standards limiting tailpipe emissions of carbon monoxide
(CO), total hydrocarbons (THC) and oxides of nitrogen
(NOx). The effect is likely to ensure that all petrol
driven cars sold within the EC after 1 January, 1993
will have to be fitted with catalytic converters. All
diesels will be required to meet these same standards,
with additional controls on particulate emissions.

To study the implications of these new Directives,
the Department of the Environment commissioned WSL to
carry out a series of vehicles studies, comparing
gaseous exhaust emissions and fuel consumption both on
the dynamometer and under real road driving conditions.
The vehicles comprised a range of 7 cars built to the
current EC Directive standard (83/351/EEC), 7 three-way
catalyst (TWC) controlled cars and 6 indirect injection
(IDI) diesel engined cars. Results from a lean-burn
car, a direct injection (DI) diesel car and an LPG car
are also included. All but two of the vehicles were
taken from the current European vehicle population and
tested in their "as-received" state.

WSL is indebted to the Department of the Environment
for permission to publish this extract of the results of
this programme in advance of the main report.

2   VEHICLE SELECTION

The cars selected for test can be grouped into 5 main categories.

Current Technology vehicles.

These 7 petrol powered cars built to the current 83/351/EEC standard and represent typical vehicles in general use across Europe. For convenience these are referred to as R15.04 cars. Emission control is achieved by conventional carburation or fuel injection matching without any special after treatment devices fitted. The car range from 1.4 to 2.2 litres capacity with an average capacity of 1.7 litres and average weight of 1069kg.

Three-Way Catalyst vehicles.

All were reportedly closed loop TWC cars and are freely available on the European market. Engine capacities range from 1.1 to 2.0 litres, with an average capacity of 1.7 litres and average reference weight of 1183kg. The cars had covered substantial mileage, but the catalysts had all been replaced within a relatively short time, typically 2000-3000km, of the reported tests.

Diesel vehicles.

All 7 diesel cars are popular vehicles freely available on the UK market. Six have IDI engines ranging from 1.6 to 2.3 litres capacity, with an average capacity of 1.9 litres and average reference weight of 1134kg. One 2.0 litre turbocharged DI vehicle is also reported.

Lean-burn vehicle.

One 1.1 litre lean-burn vehicle fitted with an oxidation catalyst is compared with a similar TWC model. Both cars were manufacturer's prototypes.

LPG Vehicle.

One 2.0 litre petrol/LPG car conversion was tested. Unlike all the other vehicles, LPG conversions are usually conducted on individual vehicles after they have been registered for road use, so avoiding the Type Approval process.

Apart from the prototypes mentioned, all vehicles were obtained from a commercial hire source or on-loan from company fleets. These vehicles were not subjected to any special preparation prior to the emission tests and should therefore be representative of the general in-service fleet.

3. TEST PROCEDURES

Measurements of CO_2 and the regulated pollutants, CO, THC and NOx, and, for diesels, particulates were carried out both in the laboratory, on a chassis dynamometer, and on the road. Fuel consumption was calculated from the gaseous emissions data using the carbon balance method, this being a computed summation of measured CO_2, CO plus THC, converted into litres of fuel consumed. Measured CO_2 differs, therefore, from calculated fuel consumption by the difference in measured CO and THC emissions.

Dynamometer Tests

The dynamometer tests were carried out using the ECE Regulation 15 urban drive cycle (R15) in both the hot and cold start modes. This gave both comparable base line data for all vehicles and a measure of each vehicles compliance with the appropriate regulation limit value. Vehicles tested later in the programme were also subjected to the new EC Extra Urban Driving Cycle (EUDC).

Emission measurements were made using an approved Constant Volume Sampler (CVS) unit. A continuous gas sample at constant flow rate was taken throughout each test drive and the emissions of CO_2, CO, THC, NOx and particulates calculated in accordance with the appropriate EC Directive. Simultaneous measurements were also made using the WSL Mini-CVS system[1].

Road Tests

Road measurements were carried out at five different driving conditions - urban, suburban, rural and two motorway drives - with average speeds of 20, 40, 60, 90 and 113kph respectively. Emission measurements spread over this normal vehicle speed range enable the characteristic emission curves for each vehicle to be derived. To minimise variability due to traffic density, etc. each road test for each car were repeated

between 4 and 10 times. The results presented herewith
are the mean averages for each test condition and for
each car group.

Exhaust samples during these road tests were
obtained using the Mini-CVS system with a mini-dilution
tunnel attachment fitted for diesel particulate
determination. This instrument samples a known
proportion, typically 1 to 2%, of diluted exhaust gas
into Tedlar bags during each drive. The bag samples are
then returned to the laboratory for analysis at the end
of the test period.

On-road THC measurements are not reported for the
diesel vehicles due to the unreliability of bag sampling
with high molecular weight hydrocarbons.

4. RESULTS AND DISCUSSION

To simplify interpretation and comparison, the data
obtained from each group of cars of the same technology
class was averaged to produce a composite emissions and
fuel consumption result. Although the mean engine
capacity and mean weight for each class of vehicles are
similar, it is recognised that, statistically, the
sample size is small and therefore definitive
conclusions should be treated with caution.
Nevertheless, since the primary objective of the
programme was to establish indicative trends only, the
following comparisons are believed to meet this
criterion. Further work has been planned by DoE over
the next two years aimed at consolidating this data.

4.1 Conventional Cars

Under this heading, three classes of vehicle are
compared - current petrol powered cars built to EC
Directive 83/351/EEC standards, three-way catalyst cars
aimed at the 1993 EC Directive, 89/458/EEC, but
available in today's market, and the IDI diesel cars,
which were also built to the current 83/351/EEC
standard.

Comparison with the COP limits. The mean exhaust
emissions for each technology class are compared in
Table 1 with the applicable conformity of production
(COP) standard defined in Directive 83/351/EEC. Also
included is a comparison against the latest Directive,
89/458/EEC, which will take effect, albeit in
substantially modified form, by 1 January 1993. This

TABLE 1. - COP Comparison - Conventional Cars.

% COP Compliance level			83/351/EEC		89/458/EEC		
Technology Class	Cap cc	Wt. kg	CO	THC+NOx	CO	THC+NOx	Part
Limit -	All	<1020	70.0	23.8	22.0	5.8	1.4
(g/test)	All	<1250	80.0	25.6	22.0	5.8	1.4
Standard	1652	1069	69%	70%	224%	290%	-
TWC Cars	1716	1183	20%	17%	63%	68%	-
IDI Diesel	1870	1077	5%	18%	17%	78%	57%

latter named draft Directive is expected to effectively mandate the introduction of catalyst cars across Europe.

It can be seen that the standard carburetted cars satisfy the COP standard to which they were built by a reasonable margin, being 69% of the CO limit and 70% of the combined limit for THC+NOx. Comparing these cars with the latest EC Directive due to take effect by 1993, they are between twice and three times the proposed COP limits for both CO and THC+NOx and, as such, provide a measure of the severity of this new legislation.

It should be noted that Directive 89/458/EEC, although a published and mandatory Directive on Member States, is unlikely to ever take effect in its present form. It is currently applicable only to cars below 1.4 litres and so is being revised by the EC Commission to embrace all sizes of passenger car with emission standards of similar severity and with coincident introduction dates. In addition, it will introduce a new high speed element into the test procedure, known as the "Extra Urban Driving Cycle" (EUDC), together with other changes such as evaporative emission control and durability measures. The cycle change will automatically result in revised limit values to take account of the higher overall average speed.

The catalyst test cars, all supposedly closed-loop three way systems, easily comply with the current standards, being less than 20% of the limit value and demonstrating the efficiency of the catalyst system with respect to the regulated pollutants. When compared to the latest EC Directive (89/458/EEC) COP levels, they are again seen to comply easily, being between 63% and

68% of the new limits. Although not tabulated here, it
has also been calculated that these same cars would meet
the forthcoming Draft Directive limits with equal or
even better margins of compliance.

The IDI diesel cars also show a substantial
improvement over the R15.04 petrol cars. Even against
the 89/458/EEC COP standard, all cars for which full
data was available complied by substantial margins,
including the particulate levels. CO compliance at 17%
compares very favourably against the 63% obtained for
the catalyst cars. This is predominantly due to the
delay in catalyst "light-off" on the cold-start R15 test
and the excess fuel introduced to aid engine start,
whereas diesel emissions are relatively independent of
test condition. Once the catalyst is alight, however,
the TWC cars generally exhibit similar levels of CO
emission as produced by the diesel cars.
Dynamometer and On-Road Emissions Comparison. As
with the dynamometer tests, the individual road test
data for each car category has been averaged into
composite results which are summarised in Tables 2 to 4.

In the following discussion, it is recognised that
the sample size is limited and hence the conclusions
should be regarded as indicative rather than definitive.
Nevertheless, although the individual model range, age
and size of the vehicles are, in most cases, different,
the overall average engine size and reference weight of
each group are not dissimilar. It is therefore
considered that the observed trends are comparable. The
diesel group has a slightly higher average engine size,
but this is also a valid representation of diesel cars
as a whole since it is common practice to fit a larger
diesel unit to obtain adequate performance due to their
specific power output being less than their petrol
driven counterparts.
CO Emissions Comparison of the mean CO emissions
show the clear advantage to be gained by the use of
catalyst or diesel technology over conventional R15.04
cars. The R15.04 cars demonstrate the classic trend of
high CO levels at low speed (12.5g/km) falling to a
minimum of 2g/km at 90kph. The hot urban road test
level of 12.5g/km is closer to the cold dynamometer R15
test result than the hot test result, reflecting the
more severe acceleration rates experienced in the real
urban driving situation.

The TWC cars show some 90% CO reduction over the
R15.04 cars, with a similar achievement from the diesel

TABLE 2. - Mean Results - ECE R15.04 Cars.

R15.04 Cars	Speed kph	CO2 g/km	CO g/km	NOx g/km	THC g/km	THC+NOx g/km	Fuel l/100km
R15 cold	19	204	12.49	1.80	2.42	4.22	9.9
R15 hot	19	177	7.77	1.66	2.26	3.92	8.4
Urban	21	192	12.53	2.46	2.32	4.78	9.3
Suburb	40	148	6.57	2.56	1.63	4.19	7.0
Rural	62	127	4.18	2.34	0.92	3.26	5.8
M/way	90	122	2.17	3.01	0.53	3.54	5.4
M/way	112	159	4.04	3.86	0.62	4.49	7.1

TABLE 3. - Mean Results - TWC Cars.

TWC Cars	Speed kph	CO2 g/km	CO g/km	NOx g/km	THC g/km	THC+NOx g/km	Fuel l/100km
R15-cold	19	267	3.42	0.29	0.63	0.85	11.6
R15-hot	19	240	0.34	0.05	0.11	0.17	10.5
EUDC	46	108	0.32	0.06	0.04	0.10	4.6
R15c+EUDC	41	136	1.07	0.11	0.19	0.26	5.8
Urban	20	238	0.79	0.16	0.08	0.25	10.1
Suburban	41	178	0.52	0.09	0.07	0.16	7.6
Rural	57	149	0.39	0.08	0.04	0.11	6.3
M90	90	137	0.21	0.08	0.02	0.10	5.8
M113	110	173	0.62	0.18	0.04	0.22	7.4

TABLE 4. - Mean Results - IDI Diesel Cars.

IDI Diesel	Speed kph	CO2 g/km	CO g/km	NOx g/km	THC g/km	THC+NOx g/km	Parts g/km	Fuel l/100km
R15 cold	19	198	0.91	0.82	0.34	1.165	0.20	7.6
R15 hot	19	165	0.92	0.73	0.41	1.139	0.22	6.4
Urban	24	176	0.74	0.63			0.26	6.8
Suburb	40	114	0.37	0.35			0.25	4.5
Rural	60	111	0.35	0.39	(0.20)	(0.59)	0.17	4.3
M/way	90	117	0.26	0.40	(0.19)	(0.59)	0.19	4.3
M/way	111	118	0.24	0.38	(0.15)	(0.53)	0.13	4.5

* Results in parentheses are estimated.

cars. The diesels also demonstrate their superiority
under cold starting conditions with little change in
emission level from a cold or a hot start. The TWC
cars, however, produce CO emissions that are a factor of
10 times that exhibited when the catalyst has "lit-up"
(3.4g/km versus 0.34g/km). Nevertheless, the efficiency
of the TWC system in producing low overall CO emissions
is clearly demonstrated.

NOx Emissions. In contrast to CO, NOx emissions
from R15.04 cars show a general rising trend with speed,
reflecting the higher load factor experienced by the
vehicle as speed increases.

The TWC cars and the IDI diesel cars behave
similarly with NOx emissions as they do with CO, the
diesels showing generally better than 80% reduction, or
<0.5g/km under most driving conditions, with the TWC
cars showing c.95% reduction, or <0.1g/km.

NOx emissions from the IDI diesel cars are some four
times higher than the TWC cars with a relatively flat
profile above urban speeds at about 0.4g/km. This still
represents an 80% to 90% reduction over the R15.04 cars
and is a substantial improvement. Under hot or cold
start dynamometer conditions, the reduction is still
better than 50%.

THC Emissions. THC emissions from the R15.04 cars
show a declining trend with speed from 2.3g/km at
c.20kph to 0.5g/km at 90kph, followed by a rise to
0.6g/km up to 113kph.

THC emissions from the TWC cars are insignificant on
all tests except the R15 cold start urban cycle, and
then at a level of only 0.6g/km. This represents a
reduction of 75% from the R15.04 base, which improves to
typically 95% once the catalyst has "lit-up".

Accurate "heated flame ionisation detector (FID)"
measurements from the IDI diesel cars were confined to
the two dynamometer test points only, but with a level
of only c.0.4g/km this represents a reduction of c.85%
over the R15.04 cars. Rough calculations suggest that
THC emissions from these IDI diesels would be c.0.2g/km
at speeds above 60kph. Table 4 lists these estimates in
parentheses, and, if correct, are a substantial
improvement over the R15.04 cars at all speeds below
90kph.

Fuel Consumption and CO_2 emissions. As may be
expected, all groups exhibit the highest fuel
consumption and CO_2 at low speeds, with minimum values
in the range between 60 and 90kph.

Comparing the fuel consumption differences between
the R15.04 cars and the TWC cars, it is seen that the
penalty for using catalyst technology, with its
necessity to maintain stoichiometry at all engine
conditions, is an increase in on-road fuel consumption
of some 9% across most of the speed range, improving to
c.3% increase at maximum motorway speed. The mean
results from the dynamometer tests show a greater
penalty of c. 20%. There is no obvious explanation for
this discrepancy between dynamometer and road results,
and may be a reflection of the low sample size. Further
vehicle tests would be required to validate this
observation.

CO_2 emissions from the TWC cars depict a similar
pattern, with a c.24% increase at low road speeds,
declining to c.8.5% increase at high road speeds. The
dynamometer tests show a 31% to 36% increase from the
R15.04 base, but the above comments regarding fuel
consumption are equally applicable. The increase over
and above the fuel consumption penalty is due to the
conversion of CO and THC into CO_2 by the catalytic
action.

The IDI diesels demonstrate their renowned fuel
economy characteristics, being typically 20% to 36%
better than the conventional petrol powered R15.04 cars
and, correspondingly, having an even greater advantage
over the TWC cars by, depending on test condition, some
25% to 45%. With CO_2 emissions, the IDI diesels show
only a 15% average improvement over the conventional
petrol cars due to the 12% greater density of diesel
fuel. The margin shown by the IDI diesels may be
reduced further if power-to-weight effects are
considered in more detail. The mean engine size of the
diesels tested is only 1870cc compared with 1652cc for
the R15.04 cars and 1716cc for the TWC cars. Published
vehicle power data suggest that in this size range the
diesel engine capacity would need to be some 400cc
larger than an equivalent petrol engine to produce
equivalent power.
      Particulate Emissions. From the overall assessment,
IDI diesel technology compares favourably with the TWC
cars on the regulated gaseous pollutants and
significantly better than all the petrol cars on fuel
economy and CO_2. Its main disadvantage, from an
emissions viewpoint, is in the level of particulates
emitted.

Table 4 shows that mean particulate emissions range

from 0.26g/km down to 0.13g/km, generally decreasing
with vehicle speed. With diesel vehicles accounting for
only 2.2% of the current UK light duty fleet this may
not constitute a major environmental problem as yet, but
a substantial increase without suitable controls on
particulate emissions could exacerbate public opinion.

4.2 Direct Injection Diesel Technology

One new technology to have been the subject of
considerable research in recent years has been the
development of the direct injection diesel car, with its
potential for even further fuel economy benefits over
its indirect injection counterpart. One DI diesel car
has so far been tested, fitted with a 2.0 litre
turbocharged engine. The vehicle was a fleet car
typical of in-service models with no prior engine
preparatory work. For comparison purposes, the results
are compared with a naturally aspirated 2.3 litre IDI
diesel car.

Comparison with COP Standards

Comparing the two vehicles with the COP limits in
Table 5, the IDI engine is substantially below both the
CO and THC+NOx limits of 83/351/EEC, and also meets the
89/458/EEC limits including the particulate standard.
The turbocharged DI engine also complies easily with
83/351/EEC, but the high NOx emissions characteristics
for the direct injection system are reflected in the
combined THC+NOx level which is more than twice the
standard set by 89/458/EEC. This vehicle additionally
exhibits a high THC value of 1.7g/km, against the
0.6g/km of the IDI vehicle, increasing its THC+NOx level
even further.

Particulate emissions are higher on the cold start
ECE R15 urban test cycle than those measured on the IDI
car, but it is clear from the levels obtained that
additional development would be needed on both engines
to meet the reductions in limits that are being
considered by the EC Commission.

On-Road Emissions - Regulated Pollutants.

The on-road emissions, and for completeness, the
dynamometer results are tabulated in Tables 6 & 7.
    CO emissions. There is little difference in CO
emissions between the two vehicle types, and the
measured levels are insignificant when compared to the

TABLE 5. - Comparison of DI and IDI Diesel Cars

DI & IDI	% COP Compliance level*				
	83/351/EEC		89/458/EEC		
Car Type	CO	THC+NOx	CO	THC+NOx	Parts
Standard -(g/test)	80.0	25.6	22.0	5.8	1.4
2.31 IDI	7%	21%	24%	95%	99%
2.01 T/DI	10%	50%	35%	220%	136%

* (Measured emissions/COP limit) x 100

TABLE 6. - Mean Results - 2.31 IDI Diesel Car.

IDI Diesel	Speed kph	CO2 g/km	CO g/km	NOx g/km	THC g/km	THC+NOx g/km	Parts g/km	Fuel l/100km
R15cold	19	250	1.32	0.80	0.55	1.35	0.34	9.7
R15 hot	19	218	1.70	0.71	0.81	1.52	0.43	8.7
Urban	20	202	0.89	0.60			0.56	7.9
Suburb	39	131	0.42	0.36	(0.17)	(0.53)	0.44	5.1
Rural	64	124	0.24	0.36	(0.04)	(0.40)	0.33	4.8
M/way	93	109	0.12	0.32	(0.04)	(0.36)	0.47	3.9
M/way	113	123	0.15	0.41	(0.03)	(0.44)	0.15	4.9

* Results in parentheses are estimated.

TABLE 7. - Mean Results - 2.01 DI Diesel Car.

IDI Diesel	Speed kph	CO2 g/km	CO g/km	NOx g/km	THC g/km	THC+NOx g/km	Parts g/km	Fuel l/100km
R15 cold	19	167	1.89	1.41	1.75	3.15	0.47	6.63
R15 hot	19	151	1.09	1.32	1.34	2.65		5.95
Urban	20	141	0.63	1.17			0.58	5.43
Suburb	43	120	0.36	1.04			0.33	4.60
Rural	59	103	0.20	0.91			0.47	3.94
M/way	95	115	0.20	1.15			0.21	4.38
M/way	112	139	0.36	1.37			0.25	5.29

R15.04 cars.

NOx Emissions. The difference in NOx emissions between the two cars is clearly illustrated, with the DI vehicle producing over twice those recorded on the IDI vehicle. This is symptomatic of the direct injection combustion system and a substantial amount of development will be needed to overcome this problem.

Particulate Emissions. A comparison of particulate emissions shows that the measured levels obtained during the road test were very similar between the two vehicles, averaging 0.56g/km at urban speeds, falling to c.0.2g/km at maximum motorway speed.

Fuel Consumption and CO_2. The DI is seen to have a 30% advantage over the IDI at urban speeds, so justifying its development. The differential, however, progressively diminishes with increasing speed and this particular example shows little difference above 40kph. It remains to be seen if this potential can be maintained with the conflicting task of reducing NOx emissions.

4.3 Lean-Burn Technology

Before the latest emissions reduction proposals were contemplated, UK vehicle manufacturers believed that the ideal compromise between emission reduction and fuel economy of small cars, < 1.4l, could be achieved with the use of "lean-burn" technology. Although substantial investment was put into this development, the latest EC Commission proposals, vis-a-vis 89/458/EEC, have overtaken the original design capability of lean-burn and no cars have reached production maturity with fully exploited emissions reduction capability. As a consequence it has not been possible to test full-production vehicles, but a 1.1 litre pre-production prototype, fitted with an oxidation catalyst, was supplied to WSL for preliminary assessment.

For comparative purposes, the lean-burn/oxycat car is compared with a standard R15.04 car and a TWC car of the same type.

Comparison with COP Standards

All 3 cars are compared with the current 83/351/EEC COP standards and the latest 89/458/EEC Directive in Table 8. The standard car exceeds the current CO limits by a wide margin, being fairly typical of general in-service vehicles, although with correct tuning for emissions rather than performance it is usually possible

TABLE 8. - Comparison with COP Standards - Lean-Burn Car

Lean-Burn	% COP Compliance level*			
	83/351/EEC		89/458/EEC	
Car Type	CO	THC+NOx	CO	THC+NOx
Standard -(g/test)	70.0	23.8	22.0	5.8
R15.04	165%	81%	525%	334%
Lean-Burn	25%	50%	80%	206%
TWC	10%	12%	32%	47%

* (Measured emissions/COP limit) x 100

to restore the CO levels to within the design standards. As was to be expected, the R15.04 car was incapable of approaching the 89/458/EEC standards.

The lean-burn car, with catalyst, met the current COP limits for both CO and THC+NOx by a 75% and 50% margin respectively, clearly demonstrating that this technology does achieve significant reductions in the regulated pollutants from typical present day levels.

This car also comfortably achieved the COP CO levels of Directive 89/458/EEC but fails to meet the THC+NOx standard by a substantial margin. The TWC car is the only vehicle of this group that complied with all the requirement of 89/458/EEC, but for its lean-burn counterpart to comply, the THC+NOx limit would need to raised twice its present level.

On-Road Emissions - Regulated Pollutants.

For CO and THC, the on-road emissions broadly reflect the dynamometer results, with the lean-burn + oxycat showing substantial reductions over the R15.04 car. No reduction in NOx emissions. however, were obtained in comparison with the standard R15.04 car.

Fuel Consumption and CO_2. The lean-burn car showed a small advantage over the R15.04 cars at road speeds at or below 40kph, but no significant difference at any other condition. The most surprising observation, however, is with the TWC car showed a slight advantage over the lean-burn car across the whole test speed range. No explanation is offered for this apparent

anomaly, but it is conceivable that the fuel economy
advantage may have been sacrificed in order to achieve
the notable improvements in CO and THC emissions.

4.4 LPG Technology

A growing number of petrol cars are now being
converted to dual-fuel operation by the addition of a
Liquid Petroleum Gas (LPG) conversion kit. The main
advantage of LPG is that the selling price is
substantially less than petrol, typically 23p/litre, due
to the lower rate of duty applied. For the high mileage
motorist the fuel price saving can quickly recover the
conversion costs.

Another marketing feature is that the LPG car is
also claimed to produce substantially less toxic
pollutants relative to the petrol car, however, as all
known conversions in the UK are carried out on
"registered" vehicles, this obviates them from the
vehicle emissions Type Approval system and so no
official figures are available.

To ascertain typical emission levels, one popular 2
litre car derived pick-up was obtained for test.

Preliminary Assessment.

As the vehicle was equipped with a simple switch to
change from LPG to petrol and vice-versa, the original
intent was to run full emission tests in both conditions
to obtain a back-to-back comparison. In the event,
preliminary trials indicated that excessive CO levels
were being produced in both conditions. It was
therefore decided to abandon the "petrol" tests as being
totally unrepresentative. With hind-sight and following
discussions with the supplier of the vehicle, this may
have been a mistake since it became clear that the high
CO from running on petrol was likely to be the direct
consequence of fitting the LPG conversion kit and,
therefore, the results would have been representative of
the real-world situation.

Regulated Emissions.

In comparison with 83/351/EEC standard, the LPG
vehicle was grossly out-of-tune with respect to CO
emissions, being some 226% of the limit value. In
contrast, THC+NOx was 26% of the limit.

The road results produced similar results at low speed, but CO levels increased to 14 times the value recorded on a similar sized petrol powered R15.04 car at 90kph. THC+NOX levels,on the other hand, were only some 60% of the levels recorded on this same petrol car. Had the system been correctly matched to the engine for optimum emissions, NOx levels would be expected to be similar to the R15.04 car.

Fuel Consumption and CO_2.

The fuel consumption and CO_2 emissions of the LPG vehicle were typically 30% to 50% higher than the comparable R15.04 car, and this must detract heavily from the benefits of the low fuel price. This may not be obvious to the vehicle driver due to the high fuel consumption that this vehicle must also have exhibited when running on petrol.

LPG Summary.

It would seem that most LPG kits are developed and marketed to suit a wide variety of engine types with the primary aim of achieving acceptable driveability and adequate fuel economy, given the low cost of the fuel. Since these kits are not necessarily developed on the same engine type to which they may ultimately be fitted, it is unlikely that optimum control of gaseous pollutants will ever be obtained. Low NOx and low THC may be obtained by default, but the penalty would appear to be excessive CO emissions. LPG should be an inherently "clean" fuel, but if this single example is typical, their full emissions reduction potential is unlikely to be realised unless full Type Approval emissions control regulations are implemented.

    6.   CONCLUSIONS

In trying to resolve the conflicting requirements of low toxic/photochemical pollutants and the reduction of the CO_2 "greenhouse" gas, both the legislators and the "environmentally aware" vehicle purchaser have a difficult task.

The EC Commission is setting controls on CO and the combined total of THC+NOx to come into effect by 1993 that will effectively require all petrol powered cars to be fitted with three-way catalyst systems. These systems have been shown to be highly efficient at reducing these regulated pollutants and, typically, will reduce tailpipe emissions by 80% to 90% from today's standards. The penalty, however, is an increase in fuel consumption and, hence, CO_2. From this limited test sample, these increases are estimated to be in the order of 9% and 16% respectively.

Conventional indirect injection diesel cars are shown to reduce the regulated pollutants by a similar order of magnitude to the catalyst cars, with CO reduced by 90% and NOx by 80%. On-road THC measurements were estimated to be significantly less than from a typical carburetted car. The main advantage shown is that average on-road fuel consumption and CO_2 emissions are reduced in the order of 29% and 15% respectively, although this benefit may reduce slightly if power requirements are taken into account. The main disadvantage of the diesel engine is its particulate emission level, being in the range of 0.26g/km to 0.13g/km.

Alternative engine technology systems have been, or are being, developed primarily aimed at low fuel costs. The single test examples discussed here, being the lean-burn petrol car, the direct injection diesel car and the LPG car all demonstrate advantages and disadvantages.

The lean-burn car with oxidation catalyst demonstrated its ability to reduce CO and THC emissions substantially from today's standard petrol cars, but NOx fuel consumption and CO_2 were relatively unaltered.

The direct injection diesel car demonstrated improved low speed fuel economy and CO_2, but suffered with the same particulate emission level as its IDI counterpart with over twice its NOx level.

The LPG car produced low THC+NOx emissions, but, due to inadequate emissions development of the carburation system, displayed excessive CO emissions and no benefit in fuel economy or CO_2 emissions reducing with vehicle speed.

7.   ACKNOWLEDGEMENTS

The author would like to express his thanks to the Department of the Environment both for its programme sponsorship and for permitting the publication of this paper in advance of the final report.

Control of Emissions from Stationary Sources of Fossil Fuel Combustion

M. J. Cooke* and R. J. Pragnell

BRITISH COAL CORPORATION, COAL RESEARCH ESTABLISHMENT, STOKE
ORCHARD, CHELTENHAM, GLOUCESTERSHIRE GL52 4RZ, UK

1 INTRODUCTION

Almost all people wish to maintain and where possible
improve their standards of living. Meeting this
objective requires extensive use of fossil fuel energy
but there has been increasing awareness that this can
have adverse effects on our environment. This is not
new. As Theodore Roosevelt said to Congress back in
1907, "to waste and to destroy our natural resources,
instead of increasing their usefulness, will undermine
the very prosperity which we are obliged to hand down to
our children, amplified and developed"(1).

Environmental pressures have been mounting
particularly over recent years. The 1980's could be
regarded as the decade in which the UK started to tackle
the Acid Rain problem. As we approach the 1990's we see
the Greenhouse Effect featuring high on the political
agenda. These pressures will not diminish. More
stringent environmental standards are being steadily
introduced. These developments represent a challenge to
industry to ensure appropriate technologies are
available for controlling emissions from fossil fuel
fired combustion plants.

In June 1988, agreement was reached in Europe on
the Large Combustion Plant Directive which specified
emission limits for SO_2, NO_x and particulates for new
plant and overall reductions for existing plant of
greater than 50 MW(t). The directive was published in
November 1988(2) and the requirements are summarised in
Table 1. Consideration is now being given in Brussels

to producing a directive to cover smaller industrial combustion plants.

Table 1   EEC Emission Standards (1988)

SO$_2$ EMISSION STANDARDS

Plant type	Solid fuel		Liquid Fuel	
	Plant size (MWt)	Emission standards (mg/m^3)	Plant size (MWt)	Emission standards (mg/m^3)
New plants	50-99	*	50-299	1700
	100	2000	300-500	***
	101-499	**		
	>500	400	7500	400
New plants firing high or variable sulphur coal	100-166	40% removal		
	167-499	§		
	>500	90% removal		

*	Limit to be decided in 1990
**	Sliding scale between 2000-400 mg/m^3
***	Sliding scale between 1700-400 mg/m^3
§	Sliding scale between 40% and 90% removal, with 60% removal at 300 MWt

NOx EMISSION STANDARDS

Plant type	Plant size (MWt)	Emission standards (mg/m^3)		
		Solid fuel	Liquid fuel	Gaseous fuel
New plants	>50	650	450	350
New plants firing coal with volatiles <10%	>50	1300		

PARTICULATE EMISSION STANDARDS

Plant type	Plant size (MWt)	Emission standards (mg/m^3)		
		Solid fuel	Liquid fuel	Gaseous fuel
New plants	50-500	100	–	–
New plants	>500	50	–	–
	All new plants	–	50	5-50*

* depending on type of gas

UK NATIONAL REDUCTION TARGETS (%)*

	1993	1998	2003
SO_2	20	40	60
NO_x	15	30	

* Reduction targets are based on 1980 as the base year.

Although it is by no means certain that serious global warming will result from man-made emissions of greenhouse gases, there is growing international consensus that precautionary measures need to be taken. General agreement was reached at the international meeting held in Toronto in 1988 that countries should reduce emissions of CO_2 from fossil fuel firing by at least 20 per cent by the year 2003 (3).

Outlined in this paper are the technical options available for controlling emissions of SO_2, NO_x, particulates and CO_2. Not all the options considered have reached the same degree of commercial maturity, but each has its advantages and disadvantages; not least there are the financial implications. And engineers in the different industrial sectors will need to examine these issues with care in order to determine the most appropriate course of action for their companies.

2 CONTROL OF SO_2 EMISSIONS

SO_2 emissions in flue gases are essentially determined

by the sulphur content of the fuel. Normally, there is negligible sulphur in natural gas whilst British coals and heavy fuel oils contain, typically, some 1-2% sulphur. A small amount of the sulphur, up to 20%, but normally about 5%, is retained in the ash when the coal is burnt.

Removal by Fuel Cleaning

Sulphur is present in fossil fuels, either as organically based sulphur (eg thiophenes) or as inorganic pyritic sulphur, FeS_2. Both coal and oil can be desulphurised by pre-combustion cleaning. During the refining process sulphur in crude oils tends to concentrate in the high boiling residual fraction. Hydrodesulphurisation processes can remove up to 90% of this sulphur. In conventional coal preparation, a substantial portion of pyritic sulphur (eg 40%) is removed as the ash content is reduced; for example from 35% ash in coal as mined to less than 10% ash in fully washed products.

The potential for removal of sulphur from coals depends on the proportion and size distribution of the inorganic pyritic sulphur which is, in principle, more easy to remove than the organic sulphur. Approximately half the sulphur from British coals is in pyritic form but is normally finely distributed throughout the coal. As a result, it is not so easily removed by physical separation methods. Nevertheless, advanced physical separation techniques, such as those based on froth flotation and High Gradient Magnetic Separation, as well as chemical and biological processes are being examined in a number of countries (20).

Removal During Combustion

Sulphur capture may be promoted by injecting dry alkaline sorbents, such as limestone, lime, etc., either into the combustion zone itself or into the flue gases just as they start to be cooled downstream. The sulphur is retained as calcium sulphate.

The effectiveness of these techniques depends on a number of factors, including choice of:

(a) Sorbent (Lime more effective but more expensive than limestone).
(b) Firing technique (Fluidised bed combustion more effective than grate or flame combustion).

(c) Operating conditions (Generally, increased sulphur capture with lower combustion temperatures).

Fluidised bed combustion (FBC) with its lower operating temperatures is particularly suited to this form of SO_2 control.

A schematic diagram of a fluidised bed boiler, with limestone addition for SO_2 control, is presented in Figure 1.

The temperature regime most favourable for sulphur capture by CaO is 800-1000°C. Higher temperatures increase the reaction rate but also lead to release of captured SO_2 by thermal decomposition of $CaSO_4$.

$$CaSO_4 \xrightarrow{>1200°C} CaO + SO_2 + \tfrac{1}{2}O_2 \qquad (1)$$

These competing influences result in an optimum temperature of about 850°C for maximum sulphur retention in FBC (see Figure 2).

Fluidised bed combustion technology, which was pioneered by British Coal, burns the coal in a bed of inert particulate material (eg coal ash, sand etc), 'fluidised' by the passage of high velocity combustion air through the bed. The intimate contact between coal and air promotes efficient combustion at relatively low temperatures, eg 900°C. When limestone sorbent is added to the bed, the same intimate contact is made between the released SO_2 and the sorbent particles at a favourable reaction temperature, ie

$$CaCO_3 \xrightarrow{>700°C} CaO + CO_2 \qquad (2)$$

$$CaO + SO_2 + \tfrac{1}{2}O_2 \xrightarrow{800-900°C} CaSO_4 \qquad (3)$$

Typically, up to 90% sulphur capture can be achieved with a variety of fluidised bed combustion systems using limestone with a Ca/S molar stoichiometry of between 2 and 3(4,5). Increases in the bed depth provide for improved utilisation of the limestone (due to better gas/solid contacting and increased sorbent residence time in the bed). This is shown in Figure 2. In contrast to FBC, with flame or grate combustion limestone is unlikely to remove more than 40% SO_2 using a similar Ca/S mole ratio; hydrated lime can increase

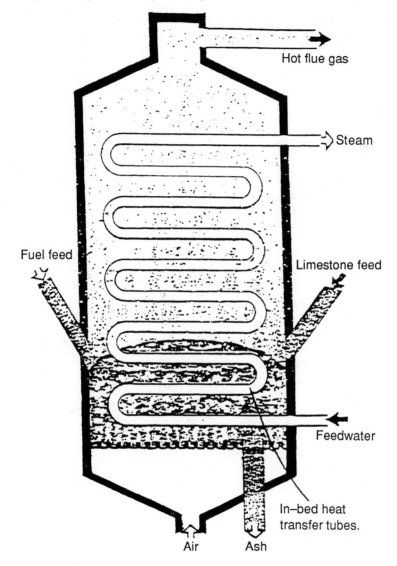

Hot flue gas

Steam

Fuel feed

Limestone feed

Feedwater

In–bed heat
transfer tubes.

Air   Ash

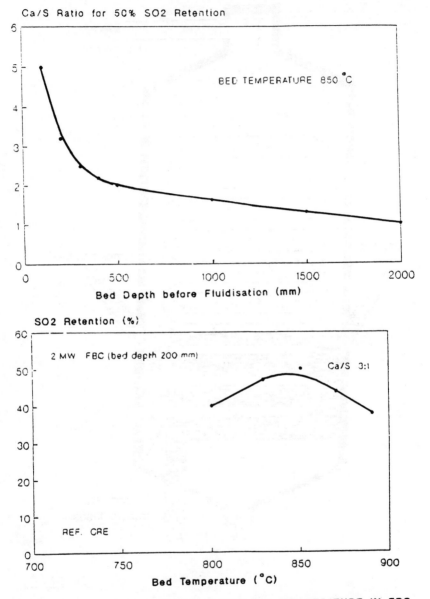

FIGURE 2.   EFFECT OF BED DEPTH AND TEMPERATURE IN FBC
ON SO2 CONTROL BY LIMESTONE ADDITION

this figure to 60%, and humidification of the combustion air can result in further improvements(6).

Flue Gas Desulphurisation

In the United Kingdom about 70% of man-made emissions of sulphur dioxide is generated from power stations(7). The most effective means of meeting the EC specific emission limits and UK SO_2 reduction targets is to fit flue gas desulphurisation (FGD) equipment to the largest fossil fuel fired power stations. The FGD plant is fitted downstream of the boiler and fly ash removal equipment where the flue gas temperature is below 250°C.

Some time ago, the CEGB announced their intention to fit all new coal-fired power stations with FGD and to retrofit 6 GW of existing generating plant. The first installation is to be at the Drax power station in North Yorkshire(8), which burns 11 million tonnes of coal per annum.

There are well over 100 FGD processes available at various stages of development, from laboratory/pilot scale through to full scale commercial operation(9). Generally, 90% SO_2 efficiencies are achieved with most FGD processes. They fall into two generic types, namely non-regenerable and regenerable systems - a classification which refers to the possible recovery of the sorbent for continued use.

Non-regenerable FGD processes are variously described as "once-through" or "throwaway". Essentially the SO_2 is absorbed in an aqueous slurry containing alkaline absorbent, typically limestone or lime.

$$Ca(OH)_2 + SO_2 \text{ -----> } CaSO_3 + H_2O \quad (4)$$

In early scrubbing systems the resulting product was a mixture of calcium sulphite and sulphate which was disposed of on the ground in spite of the solids being very difficult to dewater and forming a thixotropic sludge. In order to improve its soil mechanics properties the sludge can be stabilised by mixing it with fly ash, cement or lime.

More recently the emphasis has been towards introducing an oxidation step in which the calcium sulphite is converted to calcium sulphate in the sump of the scrubbing tower by injecting air.

$$CaSO_3 + \tfrac{1}{2}O_2 \xrightarrow{\hspace{1cm}} CaSO_4 \qquad (5)$$

In this way, high quality gypsum can be produced which can be utilised, for example, in the manufacture of plasterboard. The limestone-gypsum process has been selected by the CEGB for the Drax power station. A diagram of this method is shown in Figure 3.

With regenerable processes, the spent sorbent, typically sodium or magnesium based, is chemically or thermally regenerated for re-use whilst releasing a SO_2 rich stream which can be liquified, or used to produce sulphuric acid or elemental sulphur.

For applications involving smaller, industrial scale combustion plant, the spray-dry process (Figure 3) may be more appropriate because it is simpler and lower in capital cost(24). Instead of using a scrubbing tower, a lime slurry is atomised in a spray-dryer vessel where the water evaporates and the dry sorbent captures the SO_2 before being collected downstream in a bag filter or electrostatic precipitator. However, calcium sulphate and sulphite are produced which together with some of the unreacted lime may give rise to disposal problems.

3 CONTROL OF NOx EMISSIONS

The levels of NOx emissions in the United Kingdom are evenly divided between those derived from transport systems (motor vehicles) and those from stationary combustion sources: with respect to the latter, about 80% of NOx emissions are generated by power stations(7).

There are two sources of nitrogen oxides that result from the firing of fossil fuels; *viz* nitrogen in the fuel and atmospheric nitrogen in the combustion air. Whilst there is little nitrogen in fuel oil (less than 0.5%) and none in natural gas, coal contains some 1-2%. Depending on the method of combustion and the associated oxygen levels and temperatures, between 10 and 25% of coal nitrogen is typically converted to nitrogen oxides (fuel NOx). High temperatures lead to high conversions of fuel nitrogen, as well as promoting oxidation of atmospheric nitrogen (thermal NOx). But generally, combustion temperatures of greater than 1450°C are needed to produce significant quantities (eg 20% of total NOx with coal-firing) from the latter source.

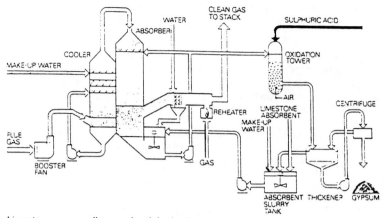

Limestone-gypsum flue gas desulphurisation process

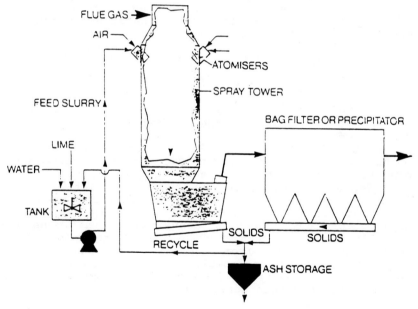

Spray Dry Desulphurisation Process

FIGURE 3. FLUE GAS DESULPHURISATION PROCESSES

Therefore, high intensity flame combustion will give the highest emissions for a particular fuel.

Typical NOx emissions from British coals by the various firing techniques are presented in Figure 4; the appropriate NOx control technology for each situation is also indicated in Figure 4. The corresponding emissions from flame combustion of natural gas and fuel oil are(22):

(i)   between 200 and 2000 mg/m³ for natural gas (some designs of gas burner operate at very high temperatures producing high levels of thermal NOx)

(ii) between 200 and 1000 mg/m³ for oil

In contrast to sulphur, which is almost all released on combustion as SO_2 and has to be captured as a secondary waste pollutant, nitrogen oxide emissions are controlled by simply converting them to nitrogen gas. This is normally achieved comparatively simply and cheaply by modifying the combustion process; but where particularly high levels of reduction are needed, the flue gases can be treated downstream of the boiler.

Combustion Modifications

There are a variety of techniques available depending, in part, on the method of coal firing; a summary of the combustion modification methods available is shown in Table 2. With high intensity flame combustion technologies, the NOx control technique is essentially the same; ie altering the distribution of fuel and air in order to encourage fuel-rich conditions during the early stages of combustion when coal volatiles are released and burnt. In this way, most of the volatile nitrogen is converted to nitrogen gas. Normally, this is achieved by diverting some of the combustion air away from the main fire zone and using the diverted air to complete the combustion of the residue char downstream from the main combustion zone. This is called staged combustion.

Table 2   NOx Control by Combustion Modification

FLAME COMBUSTION		Typical NOx Reduction (%)
1. Low NOx Burner (LNB)	– air is diverted away from the innner main combustion zone and is used to burn volatiles	30–50

in an outer secondary flame.

2.	Low Excess Air	– simply reduce and control the total air level for minimum NOx emissions and satisfactory combustion efficiency.	10-30
3.	Overfire Air (OFA)	– substantial reductions in air to the main burners with excess air being directed to ports above the burners.	20-50
4.	Reburning (Fuel Staging)	– a fraction of the fuel is injected downstream of the main flame zone followed by OFA for final combustion.	40-50 60-80 (with LNB)

FLUIDISED BED COMBUSTION

1.	Low Excess air	– reduce combustion air to 4% O_2 for low NOx emission and satisfactory combustion.	20
2.	Reduced Bed Temperature	– reduce bed temperature from 900°C to 850°C.	10
3.	Reduced Bed Particle Size	– reduce typical bed particle size from 800 μm to 600 μm.	10-20
4.	Staged Combustion	– divert up to 40% of combustion air above the bed.	30-50

Similar techniques can be used with FBC systems. In addition, though, fluidised bed combustion lends itself to other methods of NOx control. For example, operating a fluidised bed boiler at lower bed temperatures and with smaller bed particles will also encourage reductions in NOx emissions (10,11).

Sketches of staged combustion technologies for flame combustion (ie low NOx burners) and for FBC systems are presented in Figure 5.

Compliance with the EC NOx reduction targets from stationary sources(2) will require control of NOx emissions from the large fossil fuel power stations in the UK. The CEGB expect to achieve the required 30% overall reduction in NOx emissions and also meet the EC limit of 650 mg/m³ by retrofitting low NOx burners to 12 existing power stations (44 boiler units) and to all new coal-fired power stations (12).

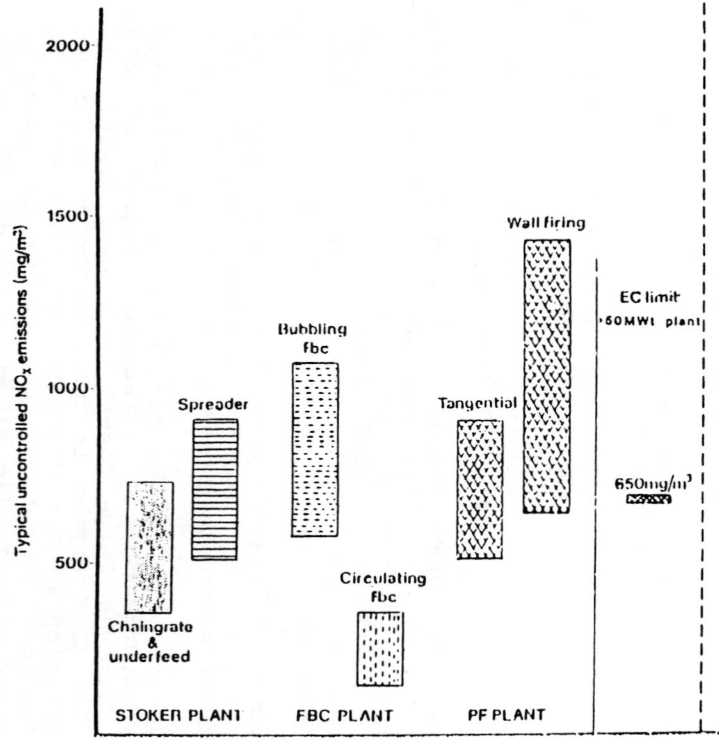

CONTROLS REQUIRED TO MEET EC LEGISLATIVE LIMITS

Boiler plant	Control target (mg/m³)	NOₓ Abatement (%)	Control technology
Chain stoker	600 → 650	nil	–
Spreader stoker	800 → 650	20	Combustion control
Tangential pf	800 → 650	20	Low NOx burner
Wall pf	1200 → 650	45	Low NOx burner
Bubbling fbc	900 → 650	35	1 Combustion control 2 Staged combustion 3 Ammonia injection
Circulating fbc	200 → 650	nil	–

FIGURE 4   NOₓ EMISSIONS FROM BRITISH COALS AND THEIR CONTROL

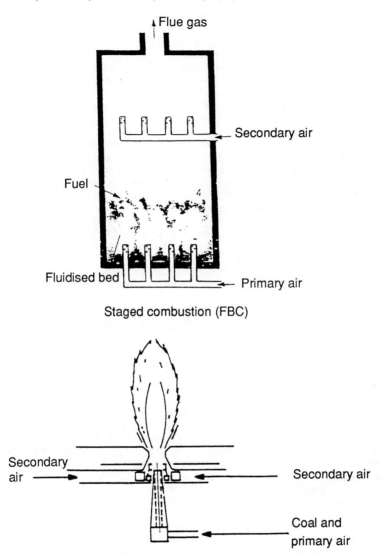

Staged combustion (FBC)

Internally staged air low no$_X$ burner (flame combustion)

238   *Energy and the Environment*

Low NOx burner trials are currently being conducted
by the power generation authorities on coal-fired
boilers. A trial on a tangentially fired boiler at
Fiddlers Ferry has demonstrated a 35-40% NOx reduction
with minimal adverse effects on boiler operation (13).
A second trial on a front wall fired boiler has
indicated a higher NOx reduction level, but in this case
there is the need for further work to determine how best
to maintain combustion efficiency (14).

In theory, higher NOx reductions of up to 50% can
be achieved and even greater reductions are possible if
more sophisticated "reburning" steps are incorporated
into new designs of combustion boilers. But care needs
to be taken to be sure that the combustion chamber can
accommodate the almost inevitably, larger combustion
zone required. For example, flame impingement can
increase the risk of corrosion to the boiler containment
tubes. Also, with lower combustion temperatures and
oxygen levels there is the danger of undesirable
hydrocarbon and carbon monoxide emissions, and reduced
combustion efficiencies unless steps are taken to avoid
this.

Chemistry of NOx Reactions during Combustion Modification

The major source of NOx emissions arising from the
combustion of coal is invariably fuel nitrogen - 95%
from FBC and 60-80% from pulverised fuel flames (15).

The nitrogenous compounds in the volatile
components of the coal are driven out during the
devolatilisation stage in the form of simple nitrogen
compounds (eg NH_3, NH, HCN). These intermediates can
then convert to NO. The nitrogen which is left in the
char particles can react with oxygen, diffusing into the
pores of the particles, to produce NO.

Although the coal volatiles and char produce NO,
they also destroy NO by reducing it to N_2. The actual
levels of NOx in the flue gas arise from the balance
between NOx formation and destruction reactions.

NO Formation

$$\text{Coal nitrogen} \xrightarrow{} \text{Simple nitrogen compounds eg } NH_3 \xrightarrow{+O_2} \text{NO} \qquad (6)$$

NO Reduction. A proportion of the NO formed is subsequently destroyed by reduction with the coal volatiles and the char.

$$NO + NX -> N_2 + OX \tag{7}$$
$$(NX = simple\ N\ compound\ eg\ NH_3)$$
$$2NO + 2C -> N_2 + 2CO \tag{8}$$
$$2NO + 2CO -> N_2 + 2CO_2 \tag{9}$$

Up to 40% of the NOx from the combustion of pulverised coal (virtually 100% in oil and gas burners) is derived from the oxidation of atmospheric nitrogen and is called thermal NOx. Thermal NOx is formed by the reaction of atmospheric nitrogen and oxygen at high temperatures.

$$N_2 + O -> NO + O \tag{10}$$
$$N + O_2 -> NO + O \tag{11}$$
$$N + OH -> NO + H \tag{12}$$

Flue Gas Treatment

If higher degrees of NOx control are required than can simply be achieved by combustion modification, it is possible to treat the flue gases either within or downstream from the boiler. The simplest approach is to inject ammonia or urea at the point where the flue gases have been cooled to about 900-1000°C. The ammonia reacts with the NO to form nitrogen gas, ie:-

$$6NO + 4NH_3 -> 5N_2 + 6H_2O \tag{13}$$
$$4NO + 4NH_3 + O_2 -> 4N_2 + 6H_2O \tag{14}$$

Relatively large amounts of ammonia need to be used for effective reduction of NO. Care needs to be taken to prevent secondary pollution, with either ammonia passing through unreacted in the flue gases or, possibly, being oxidised to form more NO.

$$4NH_3 + 5O_2 -> 4NO + 6H_2O \tag{15}$$

An early commercial process for reducing NOx by non-catalytic reaction with NH_3 was developed by Exxon and called the Thermal DeNOx process. A high NH_3:NOx stoichiometric ratio of 3:1 is required and a NOx reduction of about 70% is claimed (16). Ammonia is directly injected in the upper regions of the boiler where the narrow temperature window occurs. A major problem with this technology is the maintenance of the temperature window during load changes on the boiler. A more recent process (the NOx-Out method) uses urea as the reducing agent. The use of special additives with the urea is claimed to widen the temperature window for effective NOx reduction. Demonstration trials in Germany and Sweden indicate NOx reductions of up to 70% are possible with minimal slippage of ammonia in the stack flue gas.

Selective catalytic reduction is a more effective technique, but it is much more expensive. The flue gases are cooled down to 350-400°C and passed over a suitable catalyst (commonly vanadium pentoxide) after injecting the ammonia. NOx removals of greater than 80% are possible with minimal slippage of ammonia. Most commercial experience has been with oil-fired boilers where the flue gases are relatively free of particulates. Care needs to be taken with coal-firing to ensure that excessive dust deposition does not mask the catalyst or poison it due to the presence of alkalis and chlorides. Also, ammonia can react with SO_2 to produce amonium bisulphite in the flue gas which can cause fouling problems downstream.

Despite the higher costs, selective catalytic NOx reduction has been used for far more installations worldwide compared to the non catalytic processes.

4 COMBINED SO_2 AND NOx CONTROL

There are several processes by which SO_2 and NOx can be simultaneously removed; but they tend to be complicated and expensive technologies. In one, the flue gas is bombarded by a stream of high-energy electrons in the presence of a near stoichiometric amount of ammonia. Oxidation of both SO_2 and NOx takes place followed by a reaction with the ammonia to form ammonium sulphate and ammonium nitrate which are recovered downstream in the baghouse or electrostatic precipitator. The collected powder has potential for application as an agricultural fertiliser.

Another process has been developed where the SO_2 is adsorbed within the pores of activated carbon pellets where it is oxidised to SO_3 and reacts with water vapour to form sulphuric acid.[3] The activated char also catalyses reduction of NOx with injected ammonia. Regeneration is achieved by removing the spent carbon pellets and heating them to 650°C. The rate of attrition and poisoning of the activated carbon is a key question with regard to the viability of this process. Only a handful of installations are reported in Germany (where the technology was developed) and in Japan.

5 CONTROL OF PARTICULATE EMISSIONS

It is particularly important with coal-firing to have reliable, low cost removal of particulates from the flue gas. There are three main types of equipment available; viz cyclones, bagfilters and electrostatic precipitators. Table 3 provides a summary of their performance characteristics, (17) and Figure 6 indicates typical uncontrolled particulate emissions from British coals and the appropriate control methods used. As will be evident from the proceeding sections, particulate removal requirements are influenced by the nature of the fuel, the mode of firing and the SO_2/NOx technology being simultaneously employed.

Table 3   Characteristics   of   Particulate   Control
Technologies (Ref 17)

Characteristics	ESP	Baghouse	Cyclone/ Multicyclone
TECHNICAL Efficiency	High	High	Low-Moderate
Pressure drop	Low	Moderate	Moderate
Max operating temperature	High (up to 450°C)	Moderate (up to 260°C with conventional fabrics)	High (up to 800°C)
Sensitivity to gas temperature excursions	Performance possibly affected, but no damage	Potential damage if fabric temperature limit exceeded	Minor effects on performance
Effect of coal type	Performance best with high sulphur coal	Coal sulphur content may affect performance but less than for ESP	None-other than effect of particle size

Effect of ash type	Performance strongly influenced by ash characteristics, especially resistivity	Minor effects on performance	Abrasive ashes can cause erosion
Effect of moisture in ash	Can improve performance	Can cause caking and adversely affect performance	Can cause blocking – especially in multicyclones
Potential safety hazards	Electrical/ fire	Fire hazard if ash too hot	None
Space requirements		Moderate-high	Moderate-High Low-Moderate
Aqueous waste streams	No	No	No
ECONOMIC			
Power requirement	Low	Low-Moderate	Low-Moderate
Capital cost	High	Moderate-High	Low
Operating and Maintenance cost	Low	Moderate	Low

Where possible cyclones and multicyclones (multiple arrays of small cyclones in a single construction) are used as they provide a simple, low cost means of removing particles. Typical design features of a multicyclone are shown in Figure 7. Being inertial devices, they are relatively ineffective for collecting small (less than 5-10 micron) particles. Therefore, with increasingly stringent emission standards being introduced, the emphasis is more on the choice between bagfilters and electrostatic precipitators. Wet scrubbers are very efficient at removing particles and will tend to be favoured where there is the requirement for SO_2 removal as well. They comprise a contacting chamber, a water circuit with pumps, gas/solids settling tanks and pH control as well as giving a liquid effluent which requires suitable treatment before disposal.

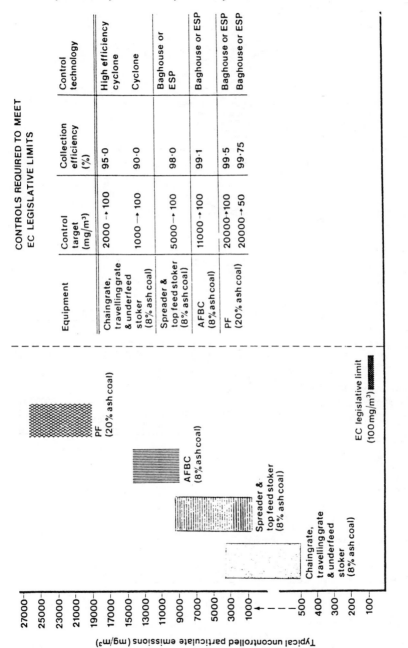

CONTROLS REQUIRED TO MEET
EC LEGISLATIVE LIMITS

Equipment	Control target (mg/m³)	Collection efficiency (%)	Control technology
Chaingrate, travelling grate & underfeed stoker (8% ash coal)	2000 → 100	95·0	High efficiency cyclone
	1000 → 100	90·0	Cyclone
Spreader & top feed stoker (8% ash coal)	5000 → 100	98·0	Baghouse or ESP
AFBC (8% ash coal)	11000 → 100	99·1	Baghouse or ESP
PF (20% ash coal)	20000 → 100	99·5	Baghouse or ESP
	20000 → 50	99·75	Baghouse or ESP

Typical unconrolled particulate emissions (mg/m³)

27000 — 25000 — 23000 — 21000 — 19000 — 17000 — 15000 — 13000 — 11000 — 9000 — 7000 — 5000 — 3000 — 1000 — 500 — 400 — 300 — 200 — 100 —

PF (20% ash coal)

AFBC (8% ash coal)

Spreader & top feed stoker (8% ash coal)

Chaingrate, travelling grate & underfeed stoker (8% ash coal)

EC legislative limit (100 mg/m³)

FIGURE 6. PARTICULATE EMISSIONS FROM BRITISH COALS AND THEIR CONTROL

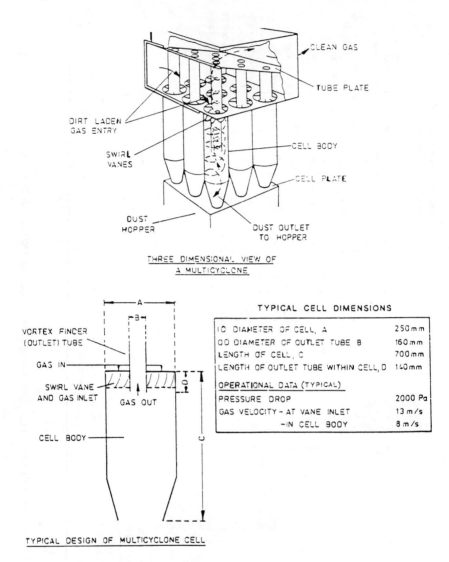

THREE DIMENSIONAL VIEW OF
A MULTICYCLONE

TYPICAL DESIGN OF MULTICYCLONE CELL

TYPICAL CELL DIMENSIONS

ID DIAMETER OF CELL, A	250 mm
OD DIAMETER OF OUTLET TUBE B	160 mm
LENGTH OF CELL, C	700 mm
LENGTH OF OUTLET TUBE WITHIN CELL, D	140 mm
OPERATIONAL DATA (TYPICAL)	
PRESSURE DROP	2000 Pa
GAS VELOCITY – AT VANE INLET	13 m/s
–IN CELL BODY	8 m/s

FIGURE 7. TYPICAL DESIGN FEATURES OF REVERSE FLOW MULTICYCLONE

Bag filters are widely used throughout the world; their popularity is reinforced by their high collection efficiency (99.5%), which includes the capture of very small particles down to less than 1 μm. A sketch of a typical bag filter design is presented in Figure 8. Bag fabric selection is crucial as the flue gas environment of high temperature and high acidity can cause premature failure after relatively short operating periods (less than 1 year). Release of the collected dust cake from the bags is commonly achieved by reverse flow or pulsed jets of compressed air applied at regular intervals. When dry sorbents have been used to capture sulphur, care needs to be taken to ensure that the bags are not damaged by a 'cementing' action of the sulphated products; this can occur if flue gas temperatures are allowed to fall below the water dewpoint in the baghouse.

Electrostatic precipitators (ESPs) have been employed frequently on large scale combustion plant such as coal-fired power stations. At this scale of operation, their high capital costs are less important than the low operating costs resulting from the low maintenance requirements and low pressure drop characteristics of ESPs. ESPs can be divided into separate cells or fields placed in series in the flue gas ductwork; this enables high collection efficiencies to be achieved. But their performance depends critically on ash resistivity. The presence of sulphur trioxide on the ash helps whereas carbon hinders collection. As a result, SO_2 and NOx control can have adverse effects on precipitator performance. Several approaches are available to overcome these adverse effects. For example, operation of the precipitator at high temperatures (ie >200°C) or alternatively, at low temperatures (ie 100°C), rather than the more usual temperature of 150°C will tend to reduce ash resistivity to optimum levels for good dust collection - see Figure 9. However, high gas temperatures result in increased flue gas volumes (larger, more expensive ESP equipment) and low temperatures increase the risks of acid condensation and corrosion. Conditioning of the flue gas by humidification can also help to lower ash resistivities to optimum levels. A diagram of an ESP is shown in Figure 8.

6 CONTROL OF CO_2 EMISSIONS

Various gases, including CO_2, CFCs, N_2O, CH_4 and O_3

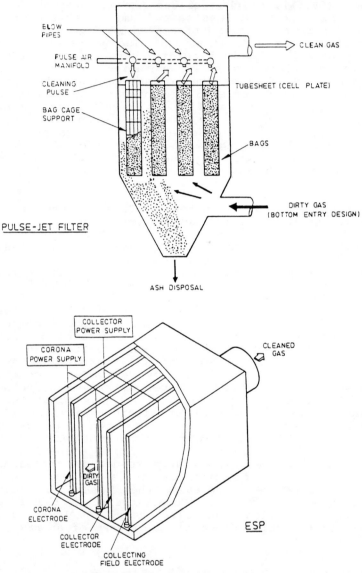

FIGURE 8 SCHEMATIC DIAGRAM OF PULSE-JET FILTER (REF. 18)
AND AN ELECTROSTATIC PRECIPITATOR

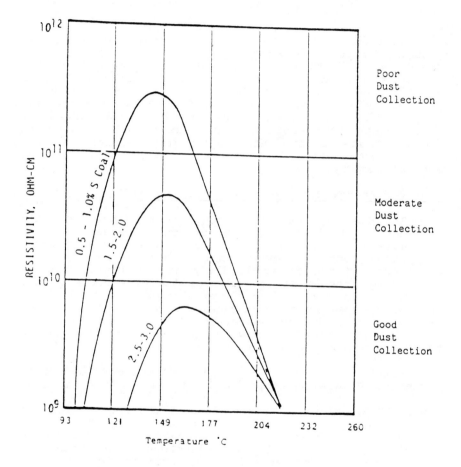

FIGURE 9. RELATIONSHIP BETWEEN ASH RESISTIVITY AND FLUE GAS
TEMPERATURE FOR COALS OF DIFFERENT SULPHUR CONTENT (REF. 23)

contribute to the Greenhouse Effect. Estimates of the present day contributions to atmospheric warming of these gases are summarised in Figure 10. CO_2 from all sources contributes some 50%. Figure 10 also gives the individual contributions of the major anthropogenic sources of CO_2, fossil fuels being responsible for about 75% of manmade CO_2 emissions.

Whilst there is no clear scientific evidence that global warming is occuring as the result of the Greenhouse Effect, it is internationally recognised that there is a need to take prompt and prudent action to counter the build up of these gases in the atmosphere. As already explained, the 1988 Toronto conference (3) recommended that CO_2 emissions should be reduced and it is envisaged that about half of this reduction could be sought from improvements in energy efficiency and conservation while the remaining half could be accomplished by fuel switching.

Improved Energy Efficiency and Conservation

The House of Commons Energy Committee have recently concluded that these two areas provide substantial scope for action (19). They have the merit of providing additional benefits beyond that of simply reducing CO_2 emissions. Since it is likely that developing nations will become the major emitters of the greenhouse gases, it is important to encourage these countries to introduce the more efficient processes for energy production used by the Western world. It is noteworthy, that the efficiency of power generation has doubled in the Western world over the past 50 years.

Further gains are obtainable if advanced combustion technologies are developed and adopted. These include combined cycle power generation and Combined Heat and Power (CHP) plant. Natural gas fired in a combined cycle can produce electricity at 50% efficiency. Advanced coal-fired systems also can give improved efficiencies and thus reductions in specific CO_2 emissions. For example, the British Coal Topping Cycle offers a 20% reduction in CO_2 emissions at a generating cost some 20% less than conventional pulverised fuel combustion with FGD (21).

CHP offers the potential of reducing CO_2 emissions as a result of doubling efficiencies by utilising the waste heat rejected in electricity generation.

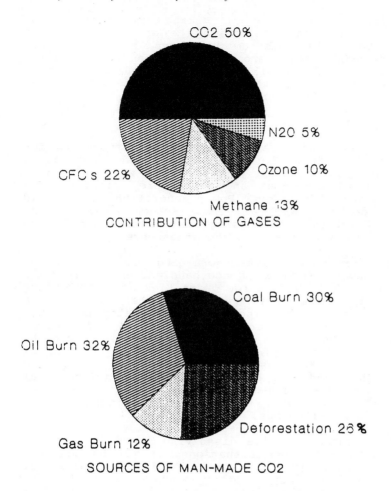

CONTRIBUTION OF GASES

SOURCES OF MAN-MADE CO2

Data obtained from various sources

FIGURE 10 CONTRIBUTIONS TO THE GREENHOUSE EFFECT

Significant progress in energy conservation was achieved in the years following the oil crises of the 1970's. But this momentum has been lost in recent years. Nevertheless, there is much scope for increasing further the standards of insulation in buildings, more efficient lighting and the substitution of products and services that are less energy intensive.

CO_2 Removal and Disposal

A number of different ways have been suggested for capturing CO_2 from flue gas discharged from power generation plant. But they all would require substantial energy input and, therefore, significantly reduce overall efficiency. Power generation costs could more than double. With all these schemes there is the underlying problem of disposing of the collected CO_2 so that it does not escape into the atmosphere.

One approach involves scrubbing the flue gases using a solvent such as monoethanolamine in order to capture up to 90% CO_2. Large quantities of low pressure steam would be needed to regenerate the solvent in a stripping column. In principle, concentrated CO_2 could then be compressed, dried, liquidified and piped offshore to deep ocean locations where it could be discharged at depths of some 3000 m to form lakes of liquid CO_2. It is assumed that the CO_2 lake would remain stable and not have any adverse environmental impact upon the sea-bed. Alternatively in some countries like the USA, the CO_2 could be deposited in solution-mined salt domes.

Other schemes involve dissolving the collected CO_2 into seawater. In some situations, it may be possible to utilise the CO_2 by pumping it down oilwells for enhanced oil recovery. Other workers advocate the approach in which the nitrogen from the combustion air is removed and CO_2 is recycled to control combustion temperatures. In this way, the CO_2 is produced already concentrated, ready for disposal.

7 CONCLUSION

As we approach the year 2000 and prepare to enter the next century, mankind is increasingly recognising that his activities can adversely affect the natural state of the world. Technologies are available and others are being developed which can effectively reduce emissions arising from fossil fuel combustion. They have differing capabilities; some technologies are more appropriate for certain applications. Their costs vary. Ultimately the consumer (ie the public) pays for pollution prevention. And one fundamental issue still needs to be faced: the extent to which the public are prepared as individuals to accept reduced standards of living and changes in lifestyles, which would inevitably result if extremely onerous environmental protection standards were to be introduced.

8 ACKNOWLEDGEMENT

This paper is published by permission of British Coal, but the views expressed are those of the authors and not necessarily of the Corporation.

9 REFERENCES

1. J. Redman, Pollution is Waste, Chemical Engineer, June 1989.

2. Council of European Communities Directive, The Limitation of Emissions of Certain Pollutants into the Air from Large Combustion Plant, 88/609/EEC, 24th November 1988.

3. Statement from the World Conference on The Changing Atmosphere, Implications for Global Security, Toronto, 27-30 June, 1988.

4. J. Makansi, B. Schwieger, Special Report on Fluidised Bed Boilers, Power, August 1982.

5. T.M. Modrak, and J.T. Tang, Sulphur Capture and Nitrogen Oxide Reduction on the 6' x 6' AFBC Test Facility, American Chem Soc (Div of Fuel Chem), Vol 27, No 1, pp 226-248, 1982.

6. B. Ekman, and R. Eloffsson, Sulphur Capture in Stoker-Fired Plant - Theory and Practical Application, IChemE Conference: Desulphurisation in Coal Combustion Systems, Sheffield, 19-21 April 1989.

7. Data from Warren Springs Laboratory, Dept of Trade and Industry 1988.

8.   CEGB Press Release, CEGB Applies to Build FGD Plant
     at Drax Power Station, January 1988.

9.   J.S. Klingspor, and D.R. Cope, FGD Handbook, Flue
     Gas Desulphurisation Systems, IEA Coal Research,
     London, 1987.

10. British Coal Final Report to the ECSC, Control of
     Nitrogen Oxides, Sulphur Oxides, Hydrocarbons and
     Particulates, ECSC Project No. 7220-ED/810,
     September 1987.

11. E. Hampartsoumian, and B. Gibbs, Journal Inst of
     Energy (402), December 1984.

12. T.F. Smith, Factors Affecting Costs of Power
     Station Emissions Control with Particular Reference
     to the UK, Enclair 1986, Taorima, Sicily, 28-31
     October 1986.

13. W. Brooks, et al, The Reduction of NOx Emissions
     from Tangentially Fired Boilers, VGB Conference,
     Strasbourg, September 1986.

14. W. Brooks, et al, Reduction of NOx Emissions from a
     500 MW Front Wall Fired Boiler, EPRI Conference,
     San Francisco, March 1989.

15. M. Hupa, Low Emission Combustion Processes
     Introductory Report, UN ECE 4th Seminar on the
     Control of Sulphur and Nitrogen Oxides from
     Stationary Sources, Graz, Austria, 12-16 May 1986.

16. G.M. Varga, and W. Bartok, Applicability of the
     Thermal DeNOx Process to Coal-Fired Utility
     Boilers, Proceedings of the Third Stationary Source
     Combustion Symp., San Francisco, March 1979.

17. P. Dacey, and D.R. Cope, Particulate Controls for
     Coal-Fired Boilers, I E A Coal Research Working
     Paper 69, London, January 1986.

18. R.C. Carr, Pulse-Jet Fabric Filters for Particulate
     Control in the US Electric Utility Industry, Proc
     of 5th Ann Int Pittsburgh Coal Conference,
     September 1988.

19. House of Commons Energy Committee, 6th Report,
     Session 1988-1989, Energy Policy Implications of
     the Greenhouse Effect, Vol 1, July 1989.

20. K. Yeager, An Assessment of Control Technology Options, 47th American Power Conference, Chicago, April 1985.

21. British Coal Brochure, Topping Cycle: Power Generation from Coal, 1989.

22. G.F. Morrison, Nitrogen Oxides from Coal Combustion - Abatement and Control, IEA Coal Research Report, No ICTIS/TR11, November 1980.

23. A.J. Buonicore, et al., Control Technology for Fine Particulate Emissions, Argonne National Laboratory Report, No. ANL/ECT-5, October 1978.

24. R.P. Bogucki, and R.J. Pragnell, Flue Gas Desulphurisation for Industrial Scale Coal-Fired Combustors - A UK Perspective, J Chem E Conference: Desulphurisation in Coal Combustion Systems, Sheffield, 19-21 April 1989.

Emission Control — Statutory Requirements

K. Speakman
HM INSPECTORATE OF POLLUTION, STOCKDALE HOUSE, VICTORIA ROAD,
LEEDS LS6 1PF, UK

1 INTRODUCTION

The control of emissions to the environment from much of
the energy producing and using industries is exercised by
H M Inspectorate of Pollution (HMIP) in England and Wales
and by H M Industrial Pollution Inspectorate in Scotland
through a number of legislative provisions. Gaseous
emissions from oil refineries, smokeless fuel plants,
fossil-fuel fired power stations etc.are controlled under
the Health and Safety at Work etc. Act 1974 and those
sections of the Alkali etc Works Regulation Act 1906 yet
to be repealed. Management of solid, liquid and gaseous
wastes from nuclear power stations and reprocessing plants
falls within the provisions of the Radioactive Substances
Act, 1960. Emissions from some of the smaller and less
complex energy using processes including domestic and
commercial heating systems are controlled by Local
Authorities under the Public Health, Clean Air and Control
of Pollution Acts. Certain aspects of the traditional
systems of pollution control operated for many years in
the UK will be changed when the Environmental Protection
Bill currently passing through Parliament receives the
Royal Assent during the Summer. These changes will also
have a significant impact on the work and operation of
HMIP and the Local Authority officers affected.

2 H M INSPECTORATE OF POLLUTION

Although HMIP was created only three years ago it brought
together Government pollution control functions, some of
which had existed for many years. In fact one of the
component parts of HMIP, the Alkali Inspectorate, later
renamed H M Industrial Air Pollution Inspectorate (HMIAPI),

has a history dating back to 1863 and was the first
industrial pollution control agency in the world. HMIP
was formed by bringing together HMIAPI, the Radiochemical
Inspectorate, the Hazardous Wastes Inspectorate and a
newly formed body charged with the control of water pollution. For the initial 2-2½ years the separate component
parts of the Inspectorate continued to operate very much
as they had previously while the plans for integration
were being developed. In October last year the first
stage of the re-organisation needed to effect the integration took place with the creation of a management
structure based on three Regions. Three Regional offices
are being created at Bristol, Bedford and Leeds under the
control of a Regional Manager who will be responsible for
all regulatory functions of HMIP in that area. Inspectors
are being organised into multi-disciplinary teams each
having responsibility for ensuring that pollution control
measures adopted by a particular group of companies under
their control are acceptable. Three Branch Heads in each
Region will co-ordinate this work under the direction of
the Regional Manager. With time, inspectors will be able
to broaden their knowledge and experience beyond their
traditional sectoral specialisations and undertake work
across the whole spectrum of HMIP's operations. However,
although the concept of an all-purpose pollution inspector
is an attractive one it is likely that some degree of
specialization will continue for many years to come.

3   POLLUTION CONTROL LEGISLATION AND ITS ENFORCEMENT

The four inspectorates from which HMIP was created are
still operating virtually independently so far as their
enforcement role is concerned and will continue to do so
until the legislative changes needed to co-ordinate
pollution control take place. The Environmental Protection Bill is introducing measures to begin this
process of integration. Emissions are currently controlled as set out below:

Emissions to Air

     The Alkali etc, Works Regulation Act 1906 and
Section 5 of the Health and Safety at Work Act 1974
require companies operating certain specified processes
to use best practicable means for the prevention of
emissions into the atmosphere and for rendering harmless
and inoffensive such emissions as may take place. The
current list of scheduled processes to which these
requirements apply together with the relevant list of
noxious and offensive substances is set out in the

Health & Safety (Emissions into the Atmosphere) Regulations 1983 and the 1989 Amendment. The list includes all
the major energy producing industries, eg power generation,
gas and coke production and petroleum and petrochemical
installations. Pollution control requirements are laid
down by HMIP, generally following discussions with repre-
sentatives of the industry concerned and checked periodically by local inspectors. Inspectors also discuss plant
modifications, investigate incidents and complaints and
advise Local Authorities on relevant air pollution matters.
In the case of new works a more open system of certifica-
tion was introduced during 1989 through The Control of Air
Pollution (Registration of Works) Regulations 1989. This
resulted from the implementation of the EEC Framework
Directive (84/360/EEC). Companies wishing to introduce
new processes must now make a formal application for permission to do so to HMIP and this application will be
made available for inspection to members of the general
public. Any comments they may make on the proposals must
be taken into account by HMIP before a certificate of
registration which allows the company to go ahead with
the development is granted.

   Best Practicable Means (BPM) requirements are updated
periodically as technology improves and when circumstances
require improvements to emission control. This has taken
place recently in the case of Large Combustion Works
(which includes fossil fired power stations) following
the formulation of measures to satisfy the EEC Large
Combustion Plant Directive (88/609/EEC). Two new sets of
BPM notes setting out requirements for plants between 50
and 500 MWth and those over 500 MWth have just been introduced. They place limits on emissions of SO_2, NO_x and
particulate matter in accordance with the requirements
of the Directive which are detailed below. The notes
apply to all new plant and to all 'substantial alterations'
to existing plants. Emissions of sulphur in the form of
SO_2 from all new plant over 500 MWth firing solid fuel
must be reduced to less than 10% of that present in the
fuel. The only practicable means of achieving reductions
on this scale will be flue-gas desulphurisation (FGD).
For boilers between 100 and 500 MWth a sliding scale of
SO_2 reductions will apply falling to 60% at 100 MWth.
The required reductions in NO_x emissions should be
achievable with the use of staged combustion in 'low-
NO_x' burners. Progressive reduction in emissions of
these pollutants in accordance with the Directive will
also require control measures to be adopted at existing
plants and a programme to meet the overall national
limits is now being formulated by HMIP.

Neither the UK Government nor the EEC has any plans yet to control CO_2 emissions.

The Large Combustion Plant Directive sets emission standards for SO_2 and NO_x for all new large combustion plant (ie 50 MW thermal input and more), and sets programmes for reducing total emissions from all existing large plants. It applies to plants burning solid, liquid or gaseous fuels but does not cover plant powered by diesel, petrol and gas engines or by gas turbines.

Member States are required to draw up programmes not later than 1 July 1990 for the progressive reduction of total annual emissions for existing plants. The programmes must set out timetables and the implementing procedures. The programmes are required to meet the emission ceilings and percentage reductions for SO_2 and oxides of nitrogen set out in Tables 1 and 2. For the UK the requirements are to reduce SO_2 emissions from existing large combustion plants to 3.11 million tonnes by 1993, 2.33 million tonnes by 1998 and 1.55 million tonnes by 2003. These figures represent reductions of 20%, 40% and 60% respectively from the 1980 baseline. The reductions in NO_x emissions required by the Directive are to 0.86 million tonnes by 1993 and 0.71 million tonnes by 1998 representing 15% and 30% reductions respectively. The programme will also have to set emission maxima for intermediate years.

From 1990 Member States are required to establish a complete emission inventory for existing plants for SO_2 and NO_x on a plant by plant basis for plants above 300 MWth and for refineries and on an overall basis for other existing plants to which the Directive applies. These inventories are already being established by HMIP in England and Wales, by HMIPI in Scotland and by the alkali and Radiochemical Inspectorate in Northern Ireland.

Article 3(4) requires the Commission to report to the Council in 1994 on the implementation of the reductions required by the Directive with, where necessary, proposals for the revision of the final SO_2 and NO_x requirements.

In addition to the controls on existing plant, Member States are also required to ensure that new (ie post 1 July 1987) plants comply with the emission limit values set out in Figures 1 and 2 and Tables 3 to 5 for SO_2, NO_x and particulates. Revised limit values are to be proposed by the Commission before 1 July 1995 in the

Table 1   Ceilings and Reduction Targets for Emissions of
          SO_2 from existing plants

Member State	1980 SO_2 emissions (ktonnes)	Emissions ceiling (ktonnes/year)		% reduction over adjusted 1980 emissions	
		Phase 1	Phase 3	Phase 1	Phase 3
		1993	2003	1993	2003
Belgium	530	318	159	-40	-70
Denmark	323	213	106	-40	-70
Germany	2225	1335	668	-40	-70
Greece	303	320	320	-45	-45
Spain	2290	2290	1440	-21	-50
France	1910	1146	573	-40	-70
Ireland	99	124	124	-29	-29
Italy	2450	1800	900	-40	-70
Luxembourg	3	1,8	1,5	-40	-50
Netherlands	299	180	90	-40	-70
Portugal	115	232	206	-25	-34
United Kingdom	3883	3106	1553	-20	-60

Ceilings for the completion of Phase 2 in 1998 are also
set out in the Directive.

light of the state of technology and environmental
requirements. Article 9 contains special rules for
setting emission limit values for new plants with multi-
fuel firing units including refineries.

    Detailed provisions for the monitoring of emissions
from new plants are contained in the Directive which also
sets out requirements for the measurement of concentrations of dust, SO_2 and NO_x. Member States are required
to notify the Commission of the total annual emissions of
SO_2 and NO_x from both new and existing large combustion
plant. To ensure that NO_x limits are met, the competent
authority (HMIP in England and Wales) may require the
adoption of appropriate design specifications and, if
monitoring reveals that the limit is not being complied
with, may require all appropriate measures to achieve
compliance.

    Member States are required to implement the Directive
by 30 June 1990 and to inform the Commission forthwith of
the means by which they have done so.

Table 2 Ceilings and Reduction Targets for Emissions of
         NO_x from existing plants

Member State	1980 NO_x emissions (as NO_2) (ktonnes)	NO_x emission ceilings (ktonnes/year)		% reduction over adjusted 1980 emissions	
		Phase 1	Phase 3	Phase 1	Phase 3
		1993	1998	1993	1998
Belgium	110	88	66	-20	-40
Denmark	124	121	81	-10	-40
Germany	870	696	522	-20	-40
Greece	36	70	70	0	0
Spain	366	368	277	-20	-40
France	400	320	240	-20	-40
Ireland	28	50	50	0	0
Italy	580	570	428	-20	-40
Luxembourg	3	2,4	1,8	-20	-40
Netherlands	122	98	73	-20	-40
Portugal	23	59	64	-8	0
United Kingdom	1016	864	711	-15	-30

Radioactive Emissions

    Control over the discharge of radioactive wastes to
the environment is the joint responsibility of the Department of the Environment (DOE) and for nuclear sites the
Ministry of Agriculture, Fisheries and Food (MAFF).
Gaseous, particulate and liquid discharges from nuclear
sites, ie those which require licensing under the Nuclear
Installations Act 1965 such as nuclear power stations,
reprocessing plants etc, are controlled by means of
Authorisations issued jointly by DOE and MAFF under the
Radioactive Substances Act, 1960 (RSA 60). Emissions of
radioactive materials from all other sites which are
registered under RSA 60 are controlled solely by HMIP.
The RSA 60 requires those who keep and use radioactive
substances to be registered with HMIP and also those
who accumulate and dispose of radioactive wastes to
obtain a Certificate fo Authorisation setting out the
conditions to be satisfied and means of disposal to be
used.

    The principal sources of emission to air from nuclear
power stations are the CO_2 purge system and in the case of

Table 3   Emission Limit Values for SO_2 for new plants -
          Gaseous fuels

Type of fuel	Limit values (mg/Nm³)
Gaseous fuels in general	35
Liquefied gas	5
Low calorific gases from gasification of refinery residues, coke oven gas, blast-furnace gas	800
Gas from gasification of coal	*

* To be decided.

Table 4   Emission Limit Values for NO_x for new plants

Type of fuel	Limit values (mg/Nm³)
Solid in general	650
Solid with less than 10% volatile compounds	1300
Liquid	450
Gaseous	350

Table 5   Emission Limit Values for Dust for new plants

Type of fuel	Thermal capacity (MW)	Emission limit values (mg/Nm³)
Solid	>500 <500	50 100
Liquid*	all plants	50
Gaseous	all plants	5 as a rule 10 for blast furnace gas 50 for gases produced by the steel industry which can be used elsewhere

* A limit value of 100mg/Nm³ may be applied to plants with a capacity of less than 500 MWth burning liquid fuel with an ash content of more than 0,06%.

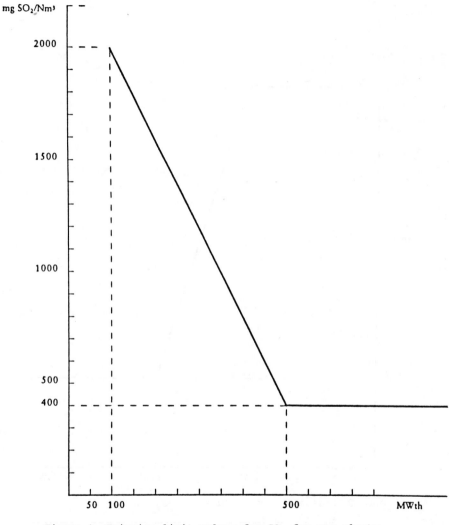

Figure 1 Emission limit values for SO_2 for new plants - Solid fuels.

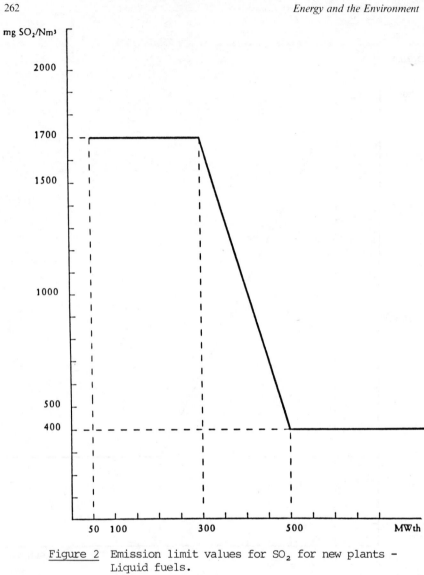

<u>Figure 2</u> Emission limit values for SO_2 for new plants -
Liquid fuels.

the older MAGNOX stations the shield cooling air while
water from the water treatment plant of the cooling ponds
and from the cooled gas conditioning system forms the
main components of the liquid effluent discharges. Low
activity solid wastes are generally sent for land disposal
at sites such as that at Drigg in Cumbria. The Government
believes that safe disposal routes for low- and intermediate-
level wastes can be developed and favour deep underground
storage facilities. UK Nirex Ltd are currently examining
a number of options for a site for a deep mine for these
materials. High-level waste will continue to be stored
at Sellafield and Dounreay until longer term options have
been fully evaluated.

Non-radioactive Liquid Discharges

Liquid waste discharges from non-nuclear energy
producing and using processes are controlled under the
provisions of the Water Act 1989, and for trade effluent
discharges to sewer, the Public Health Act 1936.
Discharges to sewers are required to meet the conditions
set by the Sewerage Undertakers based on the capacity of
the sewage tretment plant concerned. Treated effluents
from sewage works and discharges made direct to surface
waters require the consent of the pollution control
authority, now the National Rivers Authority (NRA). In
deciding on an application for consent the NRA will have
regard to the nature of the discharge and the capacity
of the receiving water to assimilate it. Discharges
have to comply with conditions set in the consent to
ensure that the relevant water quality objectives are
met. If Government proposals for creating an integrated
pollution control system are accepted discharges to both
sewer and surface waters from all industrial processes
currently under HMIP control will have to satisfy
conditions set out in the Authorisation drawn up by
HMIP. In deciding on Authorisation conditions for
discharges to water HMIP will need to ensure that water
quality objectives for the receiving water are satisfied.
They will be guided by the NRA on these matters.

Solid Wastes

The law on the control of the disposal of waste is
contained in the Control of Pollution Act 1974. This
classifies most wastes as 'controlled waste' which may
only be disposed of at a site licensed to receive that
kind of waste. There are additional controls for the
more difficult wastes, called 'special wastes' to ensure
that the movement from their point of origin can be

monitored. The responsibility for enforcing these provisions of the Act lies with the Waste Disposal Authorities (WDAs) which in most areas of England are the County Councils. In metropolitan areas the responsibilities are split between Borough Councils and single-purpose Authorities, eg the West Yorkshire Joint Waste Management Committee. In Wales, Scotland and Northern Ireland the District or Island Councils are the responsible bodies.

Waste Disposal Authorities are responsible for issuing waste disposal site licenses, preparing waste disposal plans for their areas, arranging for the disposal of much of the waste themselves, offering facilities such as civic amenity sites and monitoring special wastes.

Proposals have been published in the Environmental Protection Bill for introducing substantial changes to the law on the control of waste disposal. These include:

1. the imposition of a 'duty of care' on waste producers and holders to ensure that wastes are properly disposed of;

2. a system of registration for waste carriers;

3. powers for WDAs to take account of technical competence and financial standing in considering licences;

4. new measures against fly tipping;

5. new powers to control the import and export of wastes;

6. a requirement that WDAs transfer their waste disposal operations to arm's length companies subject to full licensing and that the WDAs be then renamed Waste Regulation Authorities;

7. strengthened guidelines for Waste Regulation Authorities, along with the requirement to publish annual reports, a statutory role for HMIP in auditing the performance of regulation authorities and certain default powers for the Secretary of State.

HMIP already audits the performance of WDAs and reports the findings to the Secretary of State. They also give advice and provide technical guidance on all waste disposal matters.

4   CHANGES IN LEGISLATION

What is arguably the most significant change to the
system of pollution control this century will take place
when the Environmental Protection Bill receives the
Royal Assent later this year. This concerns the introduction of Integrated Pollution Control (IPC). The
Royal Commission on Environmental Pollution in their
5th report in 1976 recognised that pollution can rarely
be confined to one sector of the environment and that
proper control of pollution requires a multi-media
approach. They introduced the concept of 'best
practicable environmental option ' (BPEO) and recommended
that discharges from industrial processes should be
examined as a whole. In doing so it should be possible
in many cases to identify the pattern of discharges
which causes the least damage to the environment and this
would form the basis for decisions on BPEO. If, in a
particular case, the damage caused to the environment by
liquid effluent discharges was greater than that caused
by the same quantity of pollutant discharged as a gas
from a chimney, then BPEO could be the process which
minimised the quantity of liquid effluent even if this
resulted in greater gaseous emissions than other options.
An integration of specialist knowledge in air and water
pollution and solid waste disposal would be needed to
permit BPEO decisions to be made on a rational scientific
basis. As mentioned above HMIP was created in 1987 to
work towards this integrated approach.

The main proposals for IPC contained in the Bill are:

1. all industrial processes currently under HMIP control
on account of emissions to air, together with new
categories of processes which discharge specified
dangerous substances to water (the so-called 'Red List'
substances) or which generate significant quantities of
'special wastes', will be prescribed for integrated
pollution control;

2. operators of such processes will be required to
obtain an authorisation from HMIP;

3. in granting an IPC authorisation HMIP will enforce
any existing emission limits or environmental quality
standards, require the application of 'best available
technology not entailing excessive cost' (BATNEEC) to
prevent or minimise the most polluting emissions at
source, require residual emissions to be rendered harmless to the environment, consider different disposal

options to establish what is best for the environment
as a whole and require the process to be operated
according to best practice;

4.   HMIP will consult with other bodies for environmental protection;   and

5.   applications for IPC authorisation will be advertised
and representations from the public taken into account.
Details of the consent, information on compliance and any
enforcement action taken will be held in registers to
which the public will have access.

        It should be noted that the control of radioactive
emissions will not form part of the proposed IPC system.

Acknowledgement.   Information from a number of official
Government publications has been used in the preparation
of this paper.

Renewables and Conservation

R. S. Darby

EMSTAR LIMITED, POWER CONSULTANCY DIVISION, SOUTH WING, 65
NORTHGATE, NEWARK, NOTTINGHAMSHIRE NG24 1HD, UK

1 INTRODUCTION

The first oil crisis in 1973, resulting from the Suez
war, provided fresh interest in the renewable sources of
energy and energy conservation itself. At that time
there was not only a financial incentive, but also the
prediction that energy and material resources were finite
and could be in short supply within a relatively short
period. Further these resources might be located in
hostile areas of the world which might not therefore be
readily available to all.

Energy resources have been shown to be more
extensive than this earlier prediction. Examples of this
are the availability of Russian natural gas throughout
Europe and the hydrocarbon reserves in the North Sea that
are such as to prolong the activities for longer than the
earlier forecasts. Due to the surplus of oil production
the price of a barrel of oil has fallen in real terms
since the 1970's. Nevertheless, significant improvements
in energy efficiency have taken place, against a
background in the UK of changes in the economy, switching
of fuels and an increasing demand for electricity.

The development of renewable energy has been held
back by the relatively low cost of the more traditional
forms of energy. The recent initiative by the Government
in the privatisation of electricity legislation, to
encourage the generation of electricity from this source
is to be welcomed. So far conservation has been
generally left to market forces.

The implications to the environment of atmospheric pollution and its need to reduce the gaseous products of combustion of fossil fuels required for creating primary energy has brought renewables and conservation to the forefront again. It is important to bear in mind however, that the total environmental impact of implementing these alternatives may not be entirely benign. Chemical emissions are not the only environmental facets which can effect us.

Environmental impact can be global, regional or local as is illustrated below:

Global   Regional   Local

Carbon Dioxide   Sulphur Oxides   Carbon Monoxide
Chloroflurocarbons   Nitrogen Oxides   Particles
Methane   Volatile Carbons   Heavy Metals
Nitrogen Oxide   Sound
                                                Thermal Emissions
                                                Wastes
                                                Use of Land

Further the environmental effect of each stage leading to an alternative energy system needs to be considered, before viable conclusions can be reached regarding the overall impact. This methodology is at an early stage of development and it is not yet possible to give agreed weighting to the individual environmental impacts and thus to provide reliable comparisons between the various energy sources.

2 RENEWABLES[1]

The definition of renewable energy is taken from a recent publication of the Department of Energy. Renewable energy covers those continuous energy flows that occur naturally and repeatedly in the environment - energy from the sun, the wind and the ocean and from plants and the fall of water. Biofuel produced indirectly from industrial, commercial, domestic or agricultural wastes is included. The heat from within the earth itself, geothermal energy is also usually regarded as a renewable energy source since in total it is a resource on a vast scale, although locally it cannot always maintain continuous extraction.

Estimates of present and future uses of renewable sources of energy are shown in Table 1. It can be noted

Table 1   Energy from Renewable Sources[2]

Source	Use in 1982 (kWh/year)	Utilisation in Year 2000 (kWh/year)
Solar	$2 - 3 \times 10^9$	$2 - 5 \times 10^{12}$
Geothermal	55×10^9	$1 - 5 \times 10^{12}$
Wind	2×10^9	$1 - 5 \times 10^{12}$
Tidal	0.4×10^9	$30 - 60 \times 10^9$
Wave	0	10×10^9
OTEL	0	1×10^{12}
Hydropower	1500×10^9	3×10^{12}
Biomass	$550 - 700 \times 10^9$	$2 - 5 \times 10^{12}$
Fuelwood	$10 - 12 \times 10^{12}$	$15 - 20 \times 10^{12}$
Charcoal	1×10^{12}	$2 - 5 \times 10^{12}$
TOTAL	$13 - 15 \times 10^{12}$	$31 - 53 \times 10^{12}$

UN conference on New and Renewable Sources of Energy 1981/82.

Table 2   Renewable Energy Sources

Energy Source	Liquid Fuels	Power	Heat
Solar		Thermal Electric Photovoltac Solar Power	Solar Passive
Geothermal		Turbines	Direct Heat
Wind		Turbines	Shaft Head
Hydropower		Turbines	
Ocean Energy		Tidal) Mechanical OTEC) Devices Waves)	
Biomass and Wastes	Hydro-carbons	Direct combustion of gas extracted in engines, or boilers	Direct combustion
Fuelwood & Charcoal		Direct combustion in Boilers	Direct Combustion

that the current use comes predominantly from wood fuel
in developing countries. Renewable sources of energy
provide about 15 percent of world energy supply which is
predicted to increase to 18-30 percent by the end of this
century.

The application of these renewable sources is shown
in Table 2. The first five sources, solar, geothermal,
wind, hydropower and ocean energy, create no chemical
emissions directly. The process for manufacturing the
equipment needed to convert these sources into usable
energy may do so. However, they create other
environmental problems. Wind farms are noisy and
visually intrusive. Tidal power schemes might have a
detrimental effect on the habitat of wildlife around
river estuaries and so on.

Renewables Without Chemical Emissions

The status of this range of renewable sources is as
follows:

Wind Energy. Large wind turbines (1-5 MW) have been
built in the USA and other countries with technology
approaching economic viability at favourable sites with
average annual wind speed of 5 metres per sec.   In
California there are over 18,000 wind turbines providing
1400 MW of generating capacity. There is a likelihood of
large arrays offshore before the year 2000.

Wind energy is ranked as one of the most promising
of the UK renewable sources for generating electricity.
Wind turbines have been developed and installed up to a
capacity of 3 MW and their performance is now being
monitored. Three windfarm sites each containing 25 wind
turbines in the 300-500 kW range are planned. The UK
potential for onshore generation is estimated to be 45 TW
per annum and for offshore generation 140 TW per annum.

Geothermal. This can be a commercially viable source
of power with plant sizes ranging from 1 to 110 MW.   It
is being actively investigated and developed in 30
countries. Geothermal energy can arise from underground
deposits of hot water in porous rocks and from hot dry
rocks. The programme for identifying the location of hot
water aquifers in the UK has been curtailed as there is
little prospect of them being exploited as a heat source
on their own. The UK hot dry rock programme is
proceeding, but is still at the experimental phase.

Solar Energy. Solar energy falls into three categories - solar photovoltaic cells, solar thermal electric and solar passive energy.

Photovoltaic cells have a well proven usefulness for supplying small power requirements in remote locations. However their costs would have to fall by a factor of 10 to 20 before the technology could compete economically with other energy sources for large scale electricity generation in the UK.

Whilst solar thermal electric or mirror systems are under development with some pilot plants being built, their future is not promising, except in very selected areas of the world. The development programme in the UK has been terminated.

Solar passive energy related to both domestic and non-domestic building is one of the most economically attractive renewable energy technologies in the UK.

Hydropower. Power can be extracted from water in three main ways, through tidal, wave and hydro power devices. It is estimated that there are about 40 suitable sites world-wide for tidal energy. Government wave energy research has been stopped in the UK, but it still continues in Japan and Norway. Hydropower is a mature technology and the largest exploited resources are now in the developing countries.

The UK has an unusually favourable tidal regime. The potential energy available is equivalent to 20% of the existing electricity consumption in England and Wales. Studies of the Severn Estuary and on the Mersey are proceeding. These have potential capacity for generating electricity of 7200 MW and 600 MW respectively.

Biofuels

The biofuels derived from the remaining two sources whether combusted to provide thermal energy for direct heating purposes or combusted in engines to generate electrical energy discharge gaseous emissions which are not too dissimilar to those from fossil fuels.

The biofuels arise from crops and wastes. In the UK crops means short rotation forestry specifically grown for energy purposes. Wastes can either be dry such as domestic refuse, industrial and agricultural wastes,

including straw and forest residues. The wet wastes
include sewage, animal wastes and industrial effluents.

Most dry biofuels can be burned successfully to
produce heat for use in industry or commerce. However,
the characteristics of biofuels differ markedly from
conventional fuels. They have a lower energy density, a
different chemical composition and may contain a
considerable proportion of incombustible material. In
order to burn these fuels effectively and in an
environmentally acceptable way, either conventional
combustion equipment has to be adapted or redesigned
and/or alternatively the biofuel is pre-processed to
remove incombustible materials and produce homogeneous
and easily handled fuel.

There are three means by which these biological raw
materials can be processed to create gas or liquid fuels.
They are :

-       anaerobic digestion, which results in the production
        of a rich methane gas. This takes place naturally
        in large landfill waste disposal sites and the gas
        produced is known as landfill gas (LFG) and
        typically contains methane and carbon dioxide in
        roughly equal proportions.

-       Fermentation, in which sugar or starch is converted
        to bio-ethanol by yeast. After distillation this
        can be used as a road transport fuel, but is
        unlikely to be an economic option for the UK in the
        foreseeable future.

-       Thermal processing by which the biofuel is converted
        to gaseous or liquid fuels, such as low BTU gas,
        synthetic gas, substitute natural gas, methanol or
        synthetic gasoline. These are not yet currently
        cost-effective.

Of the various biofuels and their derivatives
considered above, LFG is widely employed in the UK.

Landfill Gas[3]

The 5 billion people on this earth produce large
quantities of waste which require disposal or treatment
if harm to the environment is to be prevented and
resources protected. The UK for example produces in
excess of 20 million tonnes of organic waste per annum.
The developed countries have massive waste disposal

Table 3 Landfill Gas Composition in the U.K.

Component	Typical Value % by Volume
Methane	63.8
Carbon Dioxide	33.6
Oxygen	0.16
Nitrogen	2.4
Hydrogen	<0.05
Carbon Monoxide	<0.001
Saturated Hydrocarbons	0.005
Unsaturated Hydrocarbons	0.009
Halogenated Compounds	0.00002
Hydrogen Sulphide	0.00002
Organosulphur Compounds	<0.00001
Alcohols	<0.00001
Others	0.00005

Table 4 Purification of Landfill Gas

Impurity to be Removed	Techniques
Halocarbon	Absorption processes using activated carbons, resins, molecular sieves and liquid.
Hydrocarbon Sulphide	The Stratford process
Carbon Dioxide	Molecular sieves
Carbon Dioxide, Water and Hydrogen Sulphide	Potassium carbonate process, Seloxol process, membranes.
Carbon Dioxide and Hydrogen Sulphide	Methanol (Kryosol) process, Alkandamine process, Birax process.

industries. In contrast third world countries generate
much less waste and recover much of what they do produce.

The developed countries either incinerate their
wastes directly or indirectly or use landfilling
techniques. Both approaches can lead to severe
environmental problems. Clearly the availability of land
favours the second approach and in the UK over 90 percent
of all waste is disposed of directly to landfill.

The bacterial decomposition process taking place in
landfill sites is complicated. A typical composition of
domestic waste is:

	%
Carbohydrates	59
Inerts	31
Proteins	2.5
Plastics	1.5

The carbohydrates from paper, cardboard etc., which
form the major components of this refuse, decompose
firstly to sugar, then to mainly acetic acid and finally
to methane and carbon dioxide. The final composition of
the gas is influenced by the content of the water and
will normally contain more than 50% methane. Typical
values for LFG composition is shown in Table 3.

The current situation is that landfill sites are,
quite unintentionally, producing large quantities of
landfill gas. Left to its own devices this gas will leak
out into the atmosphere. Lateral migration underground
of gases can cause fire or even explosions and there is a
problem of smell. Further, landfill gas contains gases
such as methane and carbon dioxide which are known to
exacerbate the greenhouse effect.

The environmental and safety hazards   arising from
the generation of landfill gas have only been properly
recognised during the last 20 years. This has led to
tighter controls of general landfill practice.
Abstraction has become a necessity in order to control
and stabilise the situation. Though not the original
intention, this does make it much easier for landfill gas
to be recognised and utilised as a source of energy fuel
with a calorific value of around 20 MJ/M^3. It has been
estimated by the Department of Energy that the UK energy
potential from existing large landfill sites is of the
order of 1.3 million tonnes of coal equivalent per annum.
There are three major outlets for LFG - direct use,

(kilns, furnaces, boilers), electricity generation (using internal combustion engines or turbines) and gas cleaning and upgrading to a higher value fuel.

In the USA environmental regulations have been introduced in some states, which place strict limits on gas emissions from landfill sites and there is concern about the emissions from gas recovery systems such as engine exhausts. A separate but related issue is the potential damage to internal combustion engines used to produce electricity and waste heat, due to the halocarbon content of the landfill gas fuel. Various processes are available to purify the gas by removal of impurities. These are summarised in Table 4.

Only a small percentage of existing plants around the world use these processes as they are expensive. So far they have not been used in the UK, which has relied on a combination of filters, knockout traps and gas drying plant. These remove moisture and particulate matter and some of the waste soluble minor gas compounds with their environmental and safety hazards. Summarising therefore, there has been considerable success in collecting the inevitable discharge of LFG, and then utilising it as a fuel for electricity generation or direct heating purposes. By this means the methane is converted into the less hazardous carbon dioxide and there is a conservation of conventional fossil fuels.

3 CONSERVATION

The main primary energy sources are fossil fuels i.e. coal, oil and gas. Typical analyses are shown in Table 5. Following combustion, carbon dioxide is discharged by an amount related to the quantity of carbon in the fuel, which could be in the range of 2.8 to 3.2 gms of CO_2 per gm of fuel. The quantity of CO_2 and water vapour for a given heat output from fossil fuel are shown in Table 6. The combustion of oil discharges only about 70 per cent of CO_2 compared with coal and that from natural gas about half as much. Carbon monoxide is produced primarily during incomplete combustion of hydrocarbon fuels. The emission of nitrogen oxides (NO_x) arise from two sources, the oxidation of nitrogen in the combustion air and the oxidation of the nitrogen contained in the fuel.

Sulphur oxide (SO_x) arises from oxidation of the sulphur in the fuel and all the sulphur is normally converted to the oxide. Sulphur occurs naturally in coal

Table 5   Typical Constituents of Fossil Fuels

Constituent (by wt, dry)	Coal (%)	Oil (%)	Natural Gas (%)
Carbon	79.9	86.0	72.5
Hydrogen	5.0	11.0	23.8
Oxygen	4.3	-	-
Nitrogen	1.9	1.7	2.8
Sulphur	1.5	0.5-2.0	-
Ash	7.4	-	-
Carbon Dioxide	-	-	0.3

Table 6   Products of Combustion

	Coal	Oil	Natural Gas
Energy per kg of fuel burnt (MJ)	29	42	53
Quantities for an Output of 1000 MJ (kg)			
Fuel needed	34	24	19
Carbon content	27	20.5	14
Hydrogen content	1.7	2.6	5
CO_2 produced	99	75	51
H_2O produced	14	21	40

This ignores contributions from other constituents

and oil, but not in natural gas.

There is no realistic means available for recovering carbon dioxide at the point of discharge. Technologies are available for reducing the other emissions, either by redesign of burners, catalytic converters or by chemical routes.

It is axiomatic that conservation leading to a reduction in the usage of fossil fuels will directly or indirectly also cause a reduction in chemical emissions. This effect on chemical emissions can be brought about by the following situations :

-       Those which arise during the mining, extraction or transport of the primary source

-       The efficiency with which the fossil fuel is combusted and the products of combustion utilised

-       The means taken, with either the thermal or electrical energy generated, to minimise the usage to avoid losses and to recycle and re-utilise waste energy streams.

-       The application of more efficient techniques particularly in the field of power generation.

Production of Energy Sources

The energy industries, coal mining, oil refining, gas and electricity production, are one of the largest energy-consuming sectors of the economy. This implies that the chemical emissions associated with, for example, the combustion of oil in a boiler, are not simply those due to the combustion process, but also include those arising in the production of the oil itself.

It has been estimated that the energy requirements for extracting crude oil, the oil rig itself and transportation represents 10 to 20 percent of the total final output. Similarly the fuels used in nuclear power stations require energy for mining of the ore, its enrichment, fuel preparation and reprocessing plants. The energy usage can amount to 11 percent of the electrical output. These figures take no account of waste disposal or of the efficiency of the power station itself.

It may also be noted that burning synthetic fuels

derived from coal produces even more carbon dioxide than
burning coal directly.

Another direct loss of methane gas to the atmosphere
can occur during fossil-fuel extraction, such as coal
mining and oil extraction and also through leaks during
the transport of natural gas. These losses may
contribute as much as 10 percent of global emissions of
methane.

Combustion of Fossil Fuels

Thermal Energy. The ideal conditions for efficient
combustion are:

-      The combustion is complete. The flue gases and any
       solid residues do not contain unburnt fuel.

-      The air flow should be the minimum for complete
       combustion.

-      As much heat as possible should be extracted from
       the products of combustion and transferred to the
       receiving medium.

These conditions should apply to such duties as
steam or hot water raising and the resulting losses are
shown in Table 7. The hot water vapour formed from the
hydrogen in the fuel is a function of the fuel itself.
Natural gas contains more hydrogen than oil which in turn
contains more hydrogen than coal. The emission of water
vapour does not represent a serious environmental hazard,
but it is a significant loss of energy values. It can be
reduced in the case of natural gas by condensation and
recovery of the latent heat, providing there is a
requirement for low grade heat. The domestic condensing
boiler utilises this principle.

The major loss is the stack loss. This depends upon
the amount of air supplied for combustion and the
temperature of the gases leaving the boiler unit heating
surfaces. There has to be sufficient excess air to
provide complete combustion. Above this there will be an
unnecessary stack loss. The exit temperature is
influenced by the dew point of the products of
combustion. That from coal or oil is higher than that
from natural gas.

Where fuels are being combusted to provide a direct
process energy input, such as for kilns or drying

Table 7   Losses from Steam Raising Plant

Losses arising from burning of a fuel in a steam or hot water boiler	Typical Percentage of Gross Calorific Value of Fuel
Moisture contained in fuel	0.1 to 1.0
Moisture formed in the burning of hydrogen	4 to 10
Heat carried away in the stack	10 to 15
Incomplete combustion of carbon, volatile hydrocarbons etc.	2 to 4
Radiation	2 to 5

Table 8   U.K. Emissions[4]

Constituent	Sector	Current Emissions (millions tonnes per annum)
CO_2	Electricity	205
	Transport	120
	Domestic	90
	Industrial	94
	Commercial/Public	33
	Total	542
SO_x	Electricity	2.06
	Others	1.14
	Total	3.74
NO_x	Electricity	0.78
	Transport	0.78
	Others	0.38
	Total	1.94
CO	All sources	5.60
Volatile Organics	All sources	2.07

processes, it is frequently not possible to operate with ideal conditions for efficient combustion. For these processes there can be constraints, such as temperature, which require excess air to be employed. It then becomes important to design for the recovery of energy values from the exit flue gases.

Mechanical Energy. The combustion of fuels in engines to provide mechanical energy is influenced by pressure and temperature constraints due to the nature of the materials employed. For gas turbines, the successful development of ceramics would lead to increased temperature and hence an increase in the cycle efficiency.

Reciprocating engines used in vehicles are a major source of emissions. They produce more than 16% of the carbon dioxide and about half the carbon monoxide and nitrogen oxides, emitted in the U.K. Substantial improvements in fuel economy have been made and can still be, which reduces carbon dioxide emissions. Reductions in nitrogen oxides, carbon monoxide and unburnt hydrocarbons are expected through the development of lean burn technology, which raises the air fuel ratio and the fitting of catalytic converters. However the latter can lower fuel economy.

Table 8 gives the existing quantities of emissions from all sources discharged in the U.K.

Utilisation of Energy

Successful efforts have been made to improve process energy utilisation and hence reduce the usage of primary fuels and their associated emissions. Innovative methods of yesterday have become standard production methods of today. Further, innovations arising from a specific process or industry have been transferred successfully to other industries. Since 1965, there has been a steady reduction of UK energy consumption based on the value of goods produced, of about 1% per annum. Nevertheless a recent regional study in the UK, initiated by the EEC, concluded that savings of 20% could still be achieved.

The conventional approach to improving energy utilisation, based on ranking of opportunities, is as follows:

Good Housekeeping - Insulation - Simple Heat Recovery - Sophisticated Technologies - Process Change

The first two stages provide high ratios of return for the investment of little or no capital, but after that increasing investment is needed.

Arguably the most important analytical technique developed during the last decade for the efficient use of energy is process integration or pinch technology. This can derive the minimum heating or cooling loads for a process and also the siting and duties of the appropriate heat exchanger system and site services in order to achieve the predicted energy input.

Power Generation

The demand for electricity continues to rise. Currently one third of global primary energy is used for electricity generation. The industrial countries have installed large power stations of typically 2000 MW capacity, either nuclear or employing coal and oil as fuel. High pressure steam is raised which is expanded through steam turbines to generate power and the vapour discharged from the steam turbine is condensed. Such power stations will have efficiencies of about 35% at site, which is reduced further by distribution losses. Alternative systems are now being developed.

Combined Cycle Plant

Improved efficiency of power generation is achieved by combusting fuel directly in a prime mover such as a gas turbine, which will generate power directly and by utilising the hot exhaust gas to raise steam in a boiler which is then expanded through a steam turbine to generate additional power. Such systems can achieve efficiencies of approaching 50%. Gas turbines need clean fuels such as natural gas or gas oil. Until recently in the UK there were restrictions on the use of gas as a fuel for large power stations, but these have now been removed.

Combined cycle plants can be economically installed singly or in multiples of 300 MW units. The capital cost per MW installed is less than that for a coal fired station and there is the opportunity of siting them closer to the customer. Aside from the environmental advantages of combusting gas rather than coal, the quantity of CO_2 is greatly reduced. A coal fired station typically produces 1000 kg of CO_2 per MWh generated, whilst that from a gas fired combined cycle system is 350 kg of CO_2 per MWh generated.

The coal industry has carried out considerable research and development to use coal as the fuel for combined cycle systems. The basis of this approach is to burn coal in a fluidised bed combustor under pressure. Steam raised in the combustor is expanded through a steam turbine to generate power. The off-gas from the combustor, after cleaning, is used to drive a gas turbine.

Combined Heat and Power Plant

The combustion of fuels either directly or indirectly in prime movers i.e. gas turbines, reciprocating engines or steam turbines, leads to the conversion of thermal energy into mechanical energy which can be utilised to generate power. The efficiency of these systems either individually or in combination lies in the range of 20% to 50%. the majority of the losses are in the form of waste heat. If this waste heat can be utilised, then the cycle efficiency may be increased up to 80% or more and this is the basis of combined heat and power plant.

Figure 1 shows two systems depicting a conventional power station and a combined heat and power plant based on gas turbines. Because of process changes and reductions in thermal energy usage, which have taken place, steam turbine plants which are associated with large steam requirements are being replaced by gas turbine and reciprocating engine based systems. The latter discharge less thermal energy for the same power generated and it is in a form which can be re-utilised more flexibly. The hot gas from a gas turbine can be around 500°C and contain 15% O_2. It can be used to raise steam, with or without burning further fuel in the exhaust, or used directly for a process duty. The characteristics of the prime movers used in combined heat and power systems are shown in Table 9.

The following example shows the environmental advantages of such systems. A factory has a power requirement of 3.5 MW and steam load of 25,000 kg per hour. The power is purchased from a conventional coal fired power station and the steam is raised from a boiler on site, with an 80% efficiency. The CO_2 emitted from these two operations will be about 10,000 kg per hour. A CHP plant providing the same power and steam demands would only emit 3800 kg per hour.

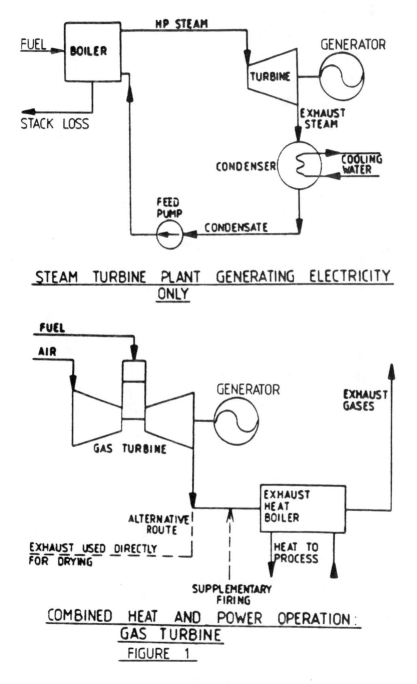

STEAM TURBINE PLANT GENERATING ELECTRICITY
ONLY

COMBINED HEAT AND POWER OPERATION:
GAS TURBINE

FIGURE 1

Table 9   Selection of CHP Systems

	Prime Mover System		
Characteristics	Gas Turbine	Reciprocating Engine	Steam Turbine
Heat to Power Ratio (H/P)	Up to 3:1	Up to 1:1	5:1 to 15:1
Flexibility in H/P by Supplementary Firing	Variable up to 15:1	Variable up to 4:1	Fixed by inlet and outlet steam pressure
Quality of Waste Heat	Gas at 450°C/ 550°C with Oxygen Content of 15%	Gas at 400°C with Oxygen Content of 12/15%, Plus Water at 80°C	Steam
Fuels Applicable	Natural Gas Gas Oil	Natural Gas Gas Oil Heavy Fuel Oil	Coal Natural Gas Gas Oil Heavy Fuel Oil

4   CONCLUSIONS

Methods are available for reducing the chemical emissions from the use of energy sources. This reduction will be off-set by the predicted increase in the requirements for energy. The implementation of these methods has so far been generally governed by market forces based on the price of energy. Recently there have been moves to prohibit or constrain various emissions. There is talk of a carbon tax which could have significant consequences.

Aside from environmental implications, some renewables have reached the stage where they are economically viable. The extraction of energy values from wastes is now an established technology. The availability and use of natural gas, which is a clean premium fuel, leads not only to reduction in chemical emissions by substitution for other fuels, but also allows the employment of much improved energy efficient systems.

References

1.   Publications by the U.K. Department of Energy 1989
2.   UN Conference on New and Renewable Sources of Energy 1981/82.
3.   Institution of Chemical Engineers Conference on Landfill Gas Production, March 1987.
4.   World Energy Yearbook 1989.

Fuel Cells and the Environment

D. S. Cameron
JOHNSON MATTHEY TECHNOLOGY CENTRE, BLOUNT'S COURT, SONNING
COMMON, READING RG4 9NH, UK

1   INTRODUCTION

Faced with a steady depletion of the world's reserves of fossil fuels, and growing opposition to nuclear power, there is an urgent need for energy generators which are both more efficient, and add a minimum of pollution to an already deteriorating environment. One exciting development – the fuel cell – meets both criteria. This is presently being developed for stationary powerplants and for transport applications. In total, several hundreds of units have been built in the USA and Japan for trials by electricity and gas supply utilities. The largest fuel cell, an 11 MW device operating on reformed natural gas, is under construction by International Fuel Cells Corporation (IFC) in the USA for installation in Tokyo.

Electric Power Generation

Recently there has been increased awareness of the pollution caused by the use of fossil fuels throughout the world. Fossil fuels (mainly coal) already account for 76% of UK electricity production, and it will be essential to minimise the effects of their combustion, if the present nuclear capacity were to be replaced.

Carbon Dioxide Emissions

The only way to reduce carbon dioxide emissions from power stations is to reduce their output, or to increase their efficiency. The latter can be achieved either by improving energy conversion and distribution processes, or by utilising the otherwise waste heat from combustion. All heat engines are limited in efficiency by the Carnot heat cycle (1), which applies equally to internal and external combustion engines. The overall efficiency of steam power generation in the United Kingdom is only 34.2%, despite having one of the most efficient generation and distribution systems in the world (2).

Power generation in the UK is based on the existence of very large centralised stations, each producing up to 2,000 MW of electrical energy and (due to the inherent inefficiency) a further 4,250 MW of low grade heat. While it is relatively easy to transport away the electrical energy generated, it is virtually impossible to utilise this heat. The low conversion efficiency means that of the total 93.8 million tonnes of coal consumed in the UK, the nation burns 61.7 million tonnes each year, not to generate electric power, but to throw away as waste heat (2). This waste represents over 220 million tonnes of carbon dioxide.

Oxides of Nitrogen and Sulphur

Most fuels, notably coal and crude oil, contain varying amounts of sulphur and nitrogen. In the case of oil, some of this may be removed during the refining process. The remainder is released on combustion of the fuel. The temperature, pressure and other conditions of the combustion process play an important part in the formation of oxides of nitrogen, and work is in progress to minimise these by switching to advanced burner designs. Efforts are being made to reduce sulphur dioxide discharges by installing scrubbers, and by changing to coal of lower sulphur content, although this is likely to result in increased costs of power generation.

Renewable Energy Options

There are a number of ways to generate power other than from fossil fuels. Electricity can be generated directly by hydroelectric schemes, wind, wave and solar energy etc. These direct energy generators will not be considered further in this paper, since it is logical to develop these as far as is practicable and economical. It is generally acknowledged that these are subject to seasonal and irregular fluctuations of output, and

limited in the amount of energy that they are able to
produce. The best estimate is that these sources could
replace the proportion of power presently generated by
nuclear energy (3).

Biomass-derived materials often represent a
disposal problem, and therefore have a negative cost as
a fuel. These include methane from land- fill sites, and
oil-field flare gases. Land-fill gases can be used efficiently by fuel cells, and there are over 1000 sites
in the UK that could be economically exploited, with a
further 400 marginally viable. The deployment of small,
unattended powerplants running on these gases could make
them highly profitable sources of energy. India, with
500 million tonnes of animal wastes available annually,
has been identified as a particularly favourable area
for their development.(4)

2   FUEL CELLS

The ideal power generator would be 100% efficient,
produce no pollutants of any kind, be completely silent,
occupy no space, and be capable of being sited at the
premises of the consumer to eliminate power transmission
lines.

There is only one type of generator which approaches these criteria in a number of respects - the fuel
cell. This is because fuel cells are electrochemical
generators, and therefore not limited by the Carnot heat
cycle. In space applications, operating on pure hydrogen
and oxygen, these have given over 70% thermal efficiency, the water produced by the fuel cells constituting
the sole source of drinking water for the crew of the
spacecraft. As electrochemical generators, they have no
moving parts except for peripheral pumps and motors, and
hence virtually no noise or vibration. They may be made
in any size from a few watts up to megawatt scale plant
with equal efficiency. Their electrical output responds
almost instantly to changing demand, while maintaining
this high efficiency, making them ideally suited to load
following applications. For terrestrial use, where
hydrogen is first generated from other fuels such as
natural gas, conversion efficiency is about 40%, with a
further 40-50% of heat available for recovery.

Although they were invented 150 years ago by a
British scientist and barrister, Sir William Grove, it

was not until the American space effort that viable devices were developed. Their potential benefits in so many
areas has led to them being adapted for terrestrial use,
mainly in combined heat and power applications, where
their attributes can be used to maximum advantage.

The US and Japanese Governments have sponsored intensive programmes so that fuel cells have now reached a
point where their feasibility has been demonstrated, and
fully engineered practical units are available for gas
and electric utility use. Productions costs have been
reduced to the point where they are commercially viable
in Japan, due to the expensive fuels, and mass production is leading to further cost reductions.Fuel cells
therefore represent an alternative means to utilise the
energy of fossil fuels in a far more efficient way than
conventional electric power generators, resulting in
less chemical, thermal and carbon dioxide emissions for
each kilowatt hour produced.

The fuel used is hydrogen or a hydrogen-rich gas
mixture, and the oxidant is normally air. The electrochemical reaction is the combination of hydrogen and
oxygen to form water. The reaction rates are enhanced by
precious metal catalysts and the devices produce both
electrical and thermal energy. (Figure 1)

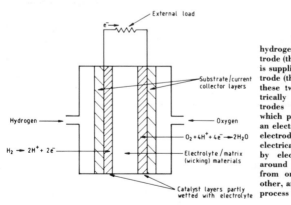

In a simple fuel cell, hydrogen is fed to one electrode (the anode) and oxygen is supplied to the other electrode (the cathode). Between these two porous and electrically conducting electrodes is the electrolyte, which permits ions to carry an electric current from one electrode to the other. The electrical circuit is completed by electrons (e⁻) flowing around the external load from one electrode to the other, and doing work in the process

<u>Figure 1</u> The simple fuel cell

There are numerous types of fuel cell in existence, characterised by the electrolyte used (Table 1). Of these, three types are being developed for combined heat and power generation. These are the phosphoric acid, the molten carbonate, and the solid oxide types. Details of their reactions, operation and construction are given elsewhere (5,6).

Phosphoric Acid Electrolyte Fuel Cells

Orthophosphoric acid (H_3PO_4) was the electrolyte chosen by Pratt and Whitney Aircraft Corporation (now part of United Technologies Corporation)in the 1960's, when they began building terrestrial fuel cells. To enable operation on natural gas or hydrocarbon fuels, a system was evolved consisting of three major sections. The first of these is a fuel processor to convert the

Types of Fuel Cells		
Type	Electrolyte	Operating temperature, °C
Alkaline	Potassium hydroxide	50-90
Proton exchange membrane	Polymeric	50-125
Phosphoric acid	Orthophosphoric acid	190-210
Molten carbonate	Lithium/potassium carbonate mixture	630-650
Solid oxide	Stabilised zirconia	900-1000
Biological	Sodium chloride	Ambient
Direct methanol	Sulphuric acid or polymer	50-120

Table 1 Types of fuel cell

fuel to hydrogen rich gas. This consists of a reformer/ water gas shift reactor. Since this is an endothermic process, some of the heat for the reaction is provided by steam from the fuel cell section, and the combustion of waste gases returned from the fuel cell.

The second component is the power section, which consists of the fuel cell stack, or multiples of stacks, fed with the hydrogen rich gas from the reformer, and atmospheric air. The reformate gases consist of a mixture of hydrogen, with some CO_2, traces (1%) of carbon monoxide, and some inert gases such as nitrogen. The fuel supply passes through the stack where about 80% of the available hydrogen is consumed. The waste gas is returned to the reformer and burned to provide heat.

The direct current from the power section may be used immediately by the consumer, or it may be converted to alternating current using a third component called an 'inverter' or power conditioner.

Phosphoric acid fuel cells operate at about 200 °C, larger scale units being pressurised at up to 8 bar. Carbon and graphite is used extensively for construction, being virtually the only inexpensive material not readily attacked by hot phosphoric acid.

Platinum and platinum/base metal alloy catalysts supported on carbon powders are used as the catalysts, which are in turn supported on carbon paper or felt electrodes, with graphite inter-cell separators. The layout of these range up to 1 m² in size, with 'stacks' of up to about 7 m high, including their outer pressure vessels producing up to 1 MW each.

The demonstrated electrical efficiency of the phosphoric acid fuel cell system is typically 38% from natural gas to bus-bar, with a further 40-45% of waste heat available from the fuel processing and the power sections together. Improvements to the reformer and system integration are likely to improve this to close to 50% in the foreseeable future.

3   DEMONSTRATION PROGRAMMES

The phosphoric acid fuel cell has been brought to an advanced state of development in a series of major programmes. The first of these was carried out by a consortium of US gas supply utilities under the TARGET (Team to Advance Research on Gas Energy Transformation) programme. As part of this, 65 fuel cells each of 12.5 kW were constructed by United Technologies Corporation (UTC) and installed and demonstrated at 35 sites in the USA, Canada and Japan by gas supply utilities.

In a second programme in the 1980's, the Gas Research Institute (GRI) and UTC developed and demonstrated a series of 40 kW powerplants, of which 46 were built. These were able to convert almost 40% of the energy in natural gas into electricity, with a further 40-45% recoverable heat.

Installation and operation of the units was carried out by various participating utilities at sites of their choice. The applications included laundries, hotels, restaurants, greenhouses, sports complexes, telephone exchanges etc..

Geographical locations varied from Alaska to Southern California, and from Tokyo to the eastern United States. Operation of the units was carefully monitored, and provided a wealth of practical experience and education. Amongst other things, these units demonstrated low NO_x emissions (less than 10% of permissible Environmental Protection Agency levels, and high reliability in unattended operation (7).

Currently, a new series of 200 kW (electrical) output CHP units are under construction by UTC and Toshiba, in a joint venture called International Fuel Cells. The first of these units are now operational, and orders for over 50 of these units have already been received. Of these, at least 10 will be coming to Europe and 25 to Japan, although none are scheduled for the UK (8).

Programmes to develop larger generators have been carried out jointly by the Electric Power Research Institute, the United States Department of Energy, and United Technologies. A 1 MW pilot plant was demonstrated in 1976, supplying electricity to the grid, and two 4.5MW generators were installed in New York by Consolidated Edison, and in Tokyo by Tokyo Electric Power Company. The latter plant was built in truck transportable modules, and was built, shipped to Tokyo and commissioned in less than 3 years. Trials of this plant have been completed.

A further 11 MW plant is currently under construction by International Fuel Cells for Tokyo Electric Power, for demonstration at the same site, and Toshiba have acquired the rights to construct further generators in Japan. This unit responds to load changes at a rate of 1 MW sec^{-1} and has a heat rate of 8,300 Btu kWh^{-1} (41.1% based on the higher heating value of the fuel) on natural gas, and will operate unattended (9). The noise

level at 100 feet from the perimeter fence is 60dB(A).
Emissions from this unit will be far lower than those
from conventional powerplants, with nitrogen oxides at 6
gGJ^{-1} and sulphur dioxide at 0.4 gGJ^{-1} compared to US
Environmental Protection Agency limits of 190 gGJ^{-1} and
800 gGJ^{-1} for gas-fired generators.

Other American companies developing phosphoric acid
fuel cells include Energy Research Corporation and Westinghouse Electric. The latter are planning a 7.5 MW powerplant using air cooling for the fuel cell stack.

Phosphoric Acid Fuel Cell Trials in Japan

Organisation	Size of cell and manufacturer
Chubu Electric Power Co.	1 MW (Toshiba/Hitachi)
Tokyo	4.5 MW (U.T.C.), 11 MW (I.F.C.), 220 kW (Sanyo), 200 kW (I.F.C.) (×2)
Kansai	1 MW (Mitsubishi/Fuji), 200 kW
Tohuku	50 kW (Fuji)
Hokkaido	100 kW (Mitsubishi)
Hokuriku	4 kW
Shikoku	4 kW (Fuji)
Okinawa	200 kW (Fuji)
Tokyo Gas Company	40 kW (U.T.C.), 200 kW (I.F.C.), 50 kW (Fuji), 100 kW (Fuji)
Osaka Gas Company	200 kW (I.F.C.), 200 kW (Mitsubishi), 50 kW/100 kW (Fuji)
Toho Gas Company	50 kW (Fuji), 100 kW (Fuji), 200 kW (I.F.C.)
Nippon Mining Co. Ltd.	100 kW (Sanyo/Toyo)

Table 2 Japanese Fuel Cell Trials

Japanese Developments of Phosphoric Acid Fuel Cells

The Japanese have organised the "Moonlight Programme" sponsored by the Ministry of International Trade
and Industry to improve the efficiency of utilisation of
fossil fuels. As part of this programme, two phosphoric
acid electrolyte fuel cell generators, each of 1 MW have
been built by two consortia of Hitachi and Toshiba, and
Mitsubishi and Fuji Electric respectively. Each of these
four companies, together with Sanyo, have separate

programmes to build phosphoric acid powerplants in sizes
ranging from a few kilowatts to 5 MW. Fuji Electric
Company have developed a 200 kW CHP unit, of which 70
have been ordered (10).

Most of the Japanese electricity generating companies are collaborating to evaluate fuel cells, as are
several of the gas supply utilities (Table 2). A complete list of industrial and utility companies participating in the Japanese programmes is published annually
with details of their efforts (11).

4   THE MOLTEN CARBONATE FUEL CELL

The molten carbonate electrolyte fuel cell (MCFC)
is at a less well developed stage than the phosphoric
acid type. The rewards offered by successfully developing
these are tremendous, although the problems associated
with working with molten salts at 650°C are considerable. The benefits include the possibility of reforming
natural gas without a separate reformer. High grade
steam is available which can be used either to run steam
turbines (a 'bottoming cycle') which raises the possible
overall efficiency to 60-65% from fuel to electric
power.

Although enormous progress has been made in developing
the MCFC up to 10 kW and even towards 100 kW in recent
years, there still remains doubt over their durability,
due to the extreme operating conditions. Tests indicate
that the system is sensitive to sulphur impurities in
the fuel stream as well as other contaminants such as
halogens, although it is envisaged that these cells will
eventually operate on coal fuel.

A number of organisations are developing MCFC's,
although none have yet offered a product for sale. In
the United States, these include UTC, Energy Research
Corporation, and the Institute of Gas Technology. In
Japan, Hitachi, Mitsubishi, Toshiba and Ishikawajima
Harima Heavy Industries are coordinated under the Moonlight programme. In Europe, the main developers are in
Holland (TNO and ECN, Petten) and in Italy, where the
programme is coordinated by the state electricity company ENEA, together with various universities, CISE
(Milan), CNR-TAE (Sicily) and Ansaldo (Milan). The
Italian objective is a 50kW MCFC unit to be developed by
1992 (12).

5   SOLID OXIDE FUEL CELLS

Solid oxide fuel cells (SOFC) have progressed in the past few years from small laboratory single cells to 3 kW powerplants, several of which have been sold for evaluation by Westinghouse, the principal developers.

Being dependent on solid state diffusion of oxygen ions through the electrolyte for conducting, the system is limited to a minimum of around 1000°C for operation. Performances have been improved by perfecting deposition methods for extremely thin layers of the zirconia electrolyte (about 40 μm) together with conducting oxide and metallic nickel electrodes.

The first system, successfully scaled up by Westinghouse, uses a system of hollow tubes connected in series and parallel by strips of nickel felt. Although the system is not compact, this design overcomes one of the fundamental problems of differential thermal expansion (13).

Alternative formats, for which power densities of up to 1 MWm^{-3} have been claimed, are dependent on packing the cells into honeycomb or flat plate type structures. These designs are being studied at Argonne National Laboratories, Ceramatec Inc., Combustion Engineering, International Fuel Cells, and Ztek Corporation in the USA,, as well as in the Government Electrotechnical Laboratory (Japan), and Dornier in Germany.

A major challenge is posed by the demands of perfectly matching the thermal characteristics of materials, which must be overcome to enable sensible scale up of the concept. At 1000°C, it would be perfectly feasible to incorporate the natural gas reforming process into the fuel cell stack, to give an extremely elegant design with high efficiency, and high grade waste heat.

Westinghouse 3 kW solid oxide fuel cells have been evaluated by the Tennessee Valley Authority, Osaka Gas Company and Tokyo Electric Power Company. Although problems have been encountered with decay after thermal cycling, a total of 12,000 hours of operation has been accumulated with the 3 units in Japan, operating on pure hydrogen, hydrogen/carbon monoxide mixtures, and reformed natural gas. A joint test is being organised by Westinghouse together with Kansai Electric Power, Osaka Gas and Tokyo Gas for a 25 kW SOFC generator. This will

incorporate internal reforming of natural gas, with
testing due to commence in 1990, thereafter progressing
to a 100 kW cogeneration system (11).

6   FUEL CELLS IN EUROPE

Efforts to develop fuel cells in Europe have been
in progress since 1985, with an EEC sponsored programme
of 10 million European Currency Units (M ecu) on solid
oxide fuel cells (SOFC), molten carbonate cells (MCFC),
direct methanol fuel cells (DMFC) and solid proton
exchange membrane fuel cells (PEMFC). It has been estimated that the total of EEC sponsorship and private fund-
ing amounts to 15 M ecu per year (12). National programmes have been started in the Netherlands, Italy and
Germany.

The first 3-year effort in The Netherlands was begun in
1986 at a cost of 9 M ecu, which has been extended to a
5-year programme (1987-1992) with an additional 25 M
ecu. This is mainly on MCFC, reflecting previous Dutch
experience of the technology, and their expertise in
materials. A small effort (2.2 M ecu) is devoted to a
demonstration of 25 kW phosphoric acid cells constructed
by Kinetics Technology International (KTI) of The Netherlands.

The Italian Government is sponsoring a national
programme (Project Volta) which will last 5 years, with
a budget of 40 M ecu divided roughly equally between
molten carbonate and phosphoric acid fuel cells. It includes constructing a 1 MW phosphoric acid powerplant
in Milan. This is being coordinated by the national
electricity supply company (ENEA), using fuel cell
stacks purchased from International Fuel Cells in the
USA, a fuel processor provided by Haldor Topsoe of
Denmark, with overall design by the Italian company
Ansaldo. In addition, smaller 25-50 kW PAFC units will
be demonstrated in Bologna and the laboratory of ENEA.
These will be constructed by KTI of The Netherlands,
with stacks supplied by Fuji Electric of Japan. In
parallel, Ansaldo are developing small (1-3 kW) PAFC
units for military applications, in collaboration with
the Italian Ministry of Defence, and ENEA. Electrodes
will be supplied by the Institute CNR-TAE, Sicily (14)
and fuel cell operating experience provided by CISE in
Milan (15).

In Sweden, there are plans to evaluate several 200 kW CHP fuel cells. Also, emphasis is being placed on the generation of hydrogen from biogas, and trials are in progress at the Royal Institute of Technology, Stockholm using a 2.5 kW fuel cell supplied by Energy Research Corporation (USA). A 25 kW unit will be the next scale. The trials are organised jointly between Swedish and Indian interests.

7   CONCLUSIONS

Fuel cells continue to make rapid strides forward, led by phosphoric technology which is now entering the market place. As with any new product, the initial pace of development has been relatively slow, with the need to educate potential customers in the electricity and gas supply industries of the safety and general viability of the devices. In general, those companies that have participated in any of the field trials have generally been keen to evaluate further examples. Although at present they are more expensive than conventional generators, the fact that they are environmentally benign will help in making them economically competitive as penalties for pollution increase. This process will also be improved by mass production bearing in mind that all the units made to date have been largely hand built, with consequent penalties in terms of price. In this respect, the lead market is likely to be in Japan, where expensive fuels make them particularly attractive. The high efficiency of these generators is matched by corresponding low pollution, which makes them uniquely suited to be the power source of the next century.

REFERENCES

1.   N.L.S. Carnot, 'Reflexions sur la Puissance Motrice du Feu', Bachelier, Paris, 1824.
2.   'Digest of United Kingdom Energy Statistics 1989', Department of Energy, H.M.S.O., London, 1989.
3.   J. Rae, 'Global Warming: UK Options and Policy' A Single European Energy Market In the Age of Environmental Awareness. January 1990, London.
4.   J.D. Pandya, 'Program and Abstracts, Fuel Cell Seminar', Tucson, Arizona, October 1986 p.361.

5. K. Kinoshita, F.R. McLarnon and E.J. Cairns, 'Fuel Cells, A Handbook" DOE/METC-88/6096, U.S. Department of Energy, Morganstown, 1988.

6. A.J. Appleby and F.R. Foulkes, 'Fuel Cell Handbook,' Van Nostrand Reinhold, New York, 1989.

7. J.M. King, R. Flacinelli and R. Woods, 'Program and Abstracts, Fuel Cell Seminar', Tucson, Arizona, October 1986 p.23.

8. F.S. Kemp and G.W. Scheffler, 'Program and Abstracts, Fuel Cell Seminar', Long Beach, California, October 1988 p.218

9. T. Ino, Y. Shiba, N. Kato and J. Caraceni. 'Program and Abstracts, Fuel Cell Seminar', Long Beach, California, October 1988 p.230

10. R. Anahara, The Grove Anniversary Symposium, London. September 1989.
    (To be published as Journal of Power Sources Parts 1 and 2, January 1990)

11. 'Fuel Cell R & D in Japan 1989', Fuel Cell Development Information Center, The Institute of Applied Energy, SY Bldg., 1-14-2, Nishinbashi, Minato-ku, Tokyo 105, Japan.

12. P. Zegers CEC-Italian Fuel Cell Workshop Proceedings June 1987, Taormina, Italy

13. S. Veyo, 'Program and Abstracts, Fuel Cell Seminar', Long Beach, California, October 1988 p.18

14. N. Giordano, E. Pasalacqua, V. Reapero, P. Staiti, V. Alderucci, R. Di Leonardo, H. Mizacan, L. Pino, & Z. Paltarzewski. 'Program and Abstracts, Fuel Cell Seminar', Long Beach, California, October 1988, p69

15. A. Ascoli, G. Redaelli. "Fuel Cells: Trends in Research and Applications" Proceedings of a Workshop Organised by UNESCO/CEC, Ravello, Italy June 1985.

Subject Index